For the Bees

Mildred Wyatt-Wold Series in
Ornithology and the Natural World

TARA DAWN CHAPMAN

For the Bees

A Handbook for Happy Beekeeping

Illustrations by Caroline Brown

UNIVERSITY OF TEXAS PRESS ⬥ AUSTIN

Requests for permission to reproduce material from this work
should be sent to permissions@utpress.utexas.edu.

♾ The paper used in this book meets the minimum requirements
of ANSI/NISO Z39.48-1992 (R1997) (Permanence of Paper).

LIBRARY OF CONGRESS CATALOGING-IN-PUBLICATION DATA

Names: Chapman, Tara Dawn, author. | Brown, Caroline (Caroline Josephine), illustrator.
Title: For the bees : a beginners guide to beekeeping / Tara Dawn Chapman ; illustrations
by Caroline Brown. Description: First edition. | Austin : University of Texas Press, 2024. |
Includes bibliographical references and index.
Identifiers: LCCN 2023051629 (print) | LCCN 2023051630 (ebook)
ISBN 978-1-4773-2951-1 (paperback)
ISBN 978-1-4773-2952-8 (pdf)
ISBN 978-1-4773-2953-5 (epub)
Subjects: LCSH: Bee culture—Handbooks, manuals, etc. Classification:
LCC SF523 .C53 2024 (print) | LCC SF523 (ebook) | DDC 638/.1—dc23/eng/20240327
LC record available at https://lccn.loc.gov/2023051629
LC ebook record available at https://lccn.loc.gov/2023051630

doi: 10.7560/329511

To Aidan, who somehow believes everything I touch turns to gold,
and convinced me this was possible. To Atlas, my favorite bee lover,
without whom this book would have been finished years earlier.
And of course for the bees, who inspire me daily.

Contents

Preface

How does one "become" a full-time beekeeper? Like many other forms of agriculture, beekeeping is a practice often passed down from one generation to the next. But not for me.

The story of how I got here is one that I don't discuss much, because talking about my past makes me a wee bit uncomfortable. When your first job out of college is one shrouded in secrecy, when all correspondence is marked with an intricate system of classifications and you are assigned a "pseudo" (short for pseudonym, or a fake name) on your first week on the job, you quickly adapt to not talking much about work.

I grew up in a blue-collar family in Smyer, Texas, a tiny town on the western side of the state with a population of just over four hundred people. I was born into the world of agriculture: like many in my area of Texas, my grandad was a cotton farmer. But I worked to get out of that tiny town as fast as I could.

After high school, I attended Duke University. I was the first in my family to attend college. A bit of a (tiny) fish out of water, I landed at Duke with eyes wide open in awe of all the world had to offer that I had yet to see. Frankly, calling it culture shock is putting it lightly. My accent was so thick, most folks did not know me as Tara but rather "Tara from Texas," because it was impossible to pull these two ideas apart from each other.

Then in 2002, in an unbelievable turn of events, I landed an interview with the Central Intelligence Agency. I was finishing my senior year at Duke and just starting to acclimate to my life there, unaware of how it was about to be flipped upside down once again. The interview was void of any expected bravado or fanfare: it was held in a low-rate hotel off the highway in Durham, North Carolina. Three months later I was offered a position, and what transpired next was eight months of personality testing, psych interviews, drug tests, background checks, and interviews with every roommate, neighbor, family member, and friend of the previous twenty-one years.

I started my new job in 2003 in a world fresh off the tragedies of 9/11, and I found myself right in the middle of the global war on terrorism. I worked in the Near East Division, which encompassed the Middle East and Southwest Asia, under what is now known as the National Clandestine Service. Over the next ten years I worked on issues related to Iraq, Pakistan, and Afghanistan from a 360-degree viewpoint. Support, intelligence gathering, policy, audits, and investigations: I touched a little of all of it.

Sometime around 2013, I was dividing my time between Austin, Texas; Washington, DC; and Kabul, Afghanistan, when I saw an ad for a discounted beekeeping class. My first thought was: "Weird. That's a thing? You can 'keep' bees?" I signed up with a woman named Gina that I met at a local coffee shop. (She would later come back full circle as a beloved Two Hives employee!)

I came home that day *obsessed* with bees. How had I lived for over thirty years and not known how incredibly fascinating these tiny insects are? I quoted bee facts nonstop to anyone who would listen. At parties, the gym, in the office, I would ramble on and on: "Did you know honey bees have these little democratic societies, and all the functions necessary to keep the colony going are performed by female bees? Did you know each colony has a queen bee, and if she dies the colony will work to produce several new queens? But did you know that the first one that emerges will race around the colony stinging all her sister queen bees so she'll be the only queen left?!" Around every corner was a new jaw-dropping honey bee biology factoid.

A few months later a new bee-obsessed friend and I built two hives together on my back deck. I am not a handy person. It wasn't pretty, and the corners didn't quite sit flush, but it was functional and served as a proper home for my first colony of bees. I couldn't remember a time when I was more proud of myself.

Just a few months after my friend and I built our first two hives and installed a few colonies of bees, I knew I wanted to start a bee business. I was fascinated that the honey from my two harvests that year each had a unique flavor and color. A bit of searching online showed me why: different flower nectars gathered by the bees create different honeys! Looking back, I realize how little I

Very few photos exist of my "prior" life. This photo was taken at Camp Joyce in Afghanistan. You can see the mountains of Pakistan in the background. Author photo.

actually knew about honey bees and beekeeping at the time. I wouldn't get a full taste of my ignorance until a few months later, but lo and behold, I was going to quit my job and start a honey company. And thus, Two Hives Honey was born.

I was in my early thirties at the time: by no means anywhere near old age, but the new life pivot certainly earned me quite a few confused looks from family and friends. I was terrified to share my plans with my coworkers in Washington, DC, worried they would question my state of mind. Interestingly enough, my DC people were more supportive than my Austin crew. Turns out a lot of government workers really just wanna buy an orchard and start a cidery. (This was a very real aspiration shared with me by a coworker in secret in a bathroom after my announcement.)

I spent the next spring season working for a commercial beekeeper out of Navasota, Texas. I got an interview, and later the job, through a cold email, which included a cover letter that cited my weightlifting personal bests and touting my experience working in war zones to demonstrate my resiliency. Four months later I returned to Austin, unsure of what the hell to do next. My "honey company" wouldn't have a real honey harvest for another year.

I lived off of savings and worked part time as a food tour guide to make ends meet. I knew I needed another way to bring in Two Hives revenue sooner. One day I was reading a business book called *The $100 Startup*, and the book advised to find a way to "get paid twice." It stuck with me. How could I get paid twice to produce a tangible jar of honey that I could only sell once?

I was struck by how in awe all my friends were anytime I let them visit my hives with me. Would people *pay* for this experience? To test my theory, I cold-called

A photo from one of my very first hive tours. We started each tour with beers and an introduction to bees on my front porch before walking across the street to see the bees. Author photo.

Meetup groups in Austin that were geared toward connecting people that love outdoor activities, pitching them my new experience. It was a hit, and "hive tours" were born: my first real product. Originally, I did these tours on six hives that a neighbor let me keep in his backyard. We began each tour on my back porch, with a cooler of cold drinks, then we would walk across the street to see the bees.

Several years later, a lot of hustle, a dash of risk-taking, and a whole lot of grit and resilience has found us where we are today: on our own property we have dubbed the Honey Ranch. It's here, just east of Austin, that we run a more sophisticated version of these early hive tours, sell our honey and other products of the hive, and host thousands of people each year interested in learning more about bees.

Like most bee businesses, we've done a little bit of everything over the years. Most beekeepers have to dabble in a few areas to make ends meet. But what we love the most is teaching about bees: on a hive tour, in a class at the Honey Ranch, or connecting online with folks all around the world. Wanting to share more about my holistic approach to beekeeping, I felt compelled to write down what I know to share with others.

This book is the product of that goal. It started in early 2021, and I wrote more than one-third of the book in four weeks. But a surprise pregnancy in February 2021 presented a few roadblocks and set my schedule back ever so slightly. It's a few years delayed, and my life looks quite different than it did when I started writing the book, but one thing hasn't changed: we are still here at the Honey Ranch every weekend, sharing our love and knowledge of bees with folks.

Working with bees has brought me so much joy and a new perspective on wildlife, ecology, and our food system. Honey bees are tricky to manage, as you will learn, but certainly not impossible. The inner workings of a honey bee colony is incredibly nuanced, and there's always more to learn. They will keep you scratching your head in curiosity, as even scientists don't completely understand their behavior and all the hows and whys of bee biology. Keep bees for the experience and you'll never be bored, but if you keep bees only for the honey, prepare to be disappointed. Enjoy the journey, and remember that all of us, no matter how long we've been in the game, still have learning to do. If you do this right, you too will remain a humble learner for life.

For the bees,

Tara Dawn Chapman

TARA DAWN CHAPMAN

For the Bees

CHAPTER 1

Honey Bee Biology

Each beekeeping journey will follow a different path. But no matter which road you travel, all beekeepers should start here, with honey bee biology.

I know where you are: I was once there too! I was *so* excited about the idea of becoming a beekeeper. Ten minutes after my first beekeeping class, I was already sketching out ideas for the logo on my honey jars and scouring the internet for fun colors of bee suits. "OH MY GOODNESS! I'm totally gonna send out a Christmas card of me in a bee suit!"

I wish someone had held up their hand and said "WHOAAA. . . ." Listen, I'm not here to rain on your honey parade. All of this planning for future honey spoils and bee suit photos and happy hour drinks in front of your hives is really fun (I give this activity four stars, and definitely recommend it, by the way), and I want you to engage in all of that. But before we get ahead of ourselves and even begin to think about purchasing bees, it's really important that you learn from my mistakes and learn a little bit about the honey bees themselves *first*. Let's jump in!

The world is home to twenty thousand species of bees, but only eight of them are considered honey bees. Of these, none are native to the Americas but were instead brought to the United States in the 1600s by colonial settlers. Honey bees belong to the order Hymenoptera, a large order of insects comprising bees, wasps, and ants, under the genus *Apis*. The most common honey bee is the western honey bee, *Apis mellifera*, which is now found on every continent except Antarctica. *A. mellifera* has dozens of different subspecies, and it is this species of honey bee that is found in the backyards of beekeepers around the world.[1]

A. Mellifera lives in large colonies consisting of thousands of individuals. This is what we call a *superorganism*, an organized society of many organisms that function as an organic whole. Additionally, honey bees are *eusocial*. Eusocial species have cooperative brood care, overlapping generations, and a reproductive division of labor that includes both sterile members and reproductive members. Other examples of eusocial superorganisms include ants, termites, and some wasps. In this social unit—similar to a really large family—each bee has a different, highly specialized job, and each bee relies on the rest of the *colony* to live. No honey bee can survive on its own for extended periods. In each of these insect colonies, the individual members are altruists and put the needs of the colony first, often at the cost of their own needs. In the honey bee world, this means that different bees are responsible for different colony tasks, from gathering food to caring for the queen to guarding the hive against intruders. Functioning as this cooperative society has allowed some species to survive (and thrive) even amid great challenges. For example, both ants and honey bees are estimated to have first appeared more than one hundred million years ago—yes, during the time of the dinosaurs!

The honey bee superorganism works as a collective unit, but within that unit they are divided into three castes, or groups. Two of the three castes in a colony, the *queen bee* and *drone bees*, are responsible only for reproduction. The third caste, the *worker bees*, are responsible for all other jobs in the colony, and they will rotate through those jobs as they age. This complex system is very nuanced, and an in-depth understanding of the biology and behavior of the colony is critical to a beekeeper's success. That's why, before we get to the "beekeeping" part, we are going to spend some time learning about bee biology. Luckily, bee biology is the most fascinating part!

Developing a holistic approach to beekeeping requires a good comprehension of the different honey bee castes and jobs, so that you understand both how they successfully meet their own needs and how and when the beekeeper should intervene and offer help. If you truly understand the needs of your honey bees and how those needs change throughout the year, you can better ensure that the bees can collect most of their needs naturally from their environment with fewer interventions from you. In the process, you will become a better beekeeper and a better bee steward. For that reason, and to best prepare you for the arrival of your bees, we will spend several chapters learning more about honey bee biology.

The Needs of a Colony

Honey bee colonies require four resources from their external environment: pollen, nectar, propolis, and water.

POLLEN

Pollen is the powdery substance you may have seen if you've looked very closely in the middle of a blooming flower. Think of pollen as the "sperm" that plants use to fertilize—or pollinate—the pistil, or ovary of the flower. This process of pollination is what allows the now-fertilized *egg* to produce a seed, ensuring the plant can produce offspring.

Pollen is very nutritious—full of vitamins, minerals, *carbohydrates, protein,* and *lipids*—and foraging honey bees collect it and carry it back to the hive on their hind legs. Beekeepers refer to this part of the bee's anatomy as the *pollen baskets.* This nutritionally rich bee superfood is essential for rearing healthy *brood,* and honey bees feed it to the young, developing bees in a colony.

Honey bees ferment the pollen when storing it, using a bit of *honey* or *nectar* to help ensure the preservation of its nutritional benefits. This fermentation process produces what we call *bee bread.* Bee bread is stored in tiny hexagons made out of *beeswax* until it is needed to feed the developing honey bees. You will learn more about this beeswax comb in chapter 2.

Far too many beekeepers misunderstand the role of pollen in a colony and mistakenly believe that bees use it to produce honey. For that, honey bees have to collect nectar.

NECTAR

This is the secret in the honey sauce! Nectar is a sugar-rich liquid produced by the "nectaries" in some flowers. Flowers produce this nectar in order to attract potential pollinators, such as honey bees, butterflies, or hummingbirds, to help ensure pollination. When a pollinator visits a flower in search of nectar, it also gets dusted in pollen. Each time a pollinator visits a new plant, some of the pollen it was carrying from the previous plants falls into the blossom, and the fertilization cycle begins.

But why are pollinators on the hunt for nectar in the first place? For many adult pollinators, including honey bees, nectar is their predominant food source. Foraging bees gather the nectar from flowers and carry it back to their colonies in a second "stomach" called a *honey crop.* They then dehydrate the nectar to produce the honey we love! You will learn more about both pollen and nectar and their roles in a honey bee's diet in chapter 3, "Honey Bee Nutrition."

PROPOLIS

Honey bees gather resins from the sap of trees and other botanical sources, carry it back to the hive in their pollen baskets, and mix it with saliva, beeswax, and honey to create what beekeepers call bee glue. Honey bees use this *propolis* glue to seal up all small cracks in the hive, fill in rough surfaces, and waterproof it from rain and the elements. And because propolis is antibacterial and antiviral, it may help protect the colony from pathogens too.[2]

Bees also use propolis to mummify intruders. If another insect (or even a mouse!) invades a hive, the bees can act quickly to kill it. But some predators are too big for tiny bees to carry back out, so they wrap the deceased intruder in propolis to keep the dead body from contaminating the colony. Finally, honey bees line the beeswax cells with propolis before each generation of eggs is laid in the comb to help protect the developing brood from disease. Propolis can be harvested by the beekeeper and is used in formulations for colds, treatments for burns and wounds, and toothpastes and mouthwashes to treat gingivitis.

WATER

Honey bees use water in two ways. First, they use it to dilute honey so that it can be fed more easily to developing honey bees. And second, they use water to create an evaporative cooling system inside the hive during the hottest summer months: think of this as honey bee air conditioning!

The Caste of Characters

Before we dive into the nuances of managing honey bee colonies, let's meet the "caste" of characters in a colony! Within a honey bee colony live three different castes of bees:

1. A queen bee (a fertile female)
2. Worker bees (infertile females)
3. Drone bees (males)

The development cycle for each of these castes is the same. They will each start out as a tiny egg, then hatch into a *larva*, develop into a *pupa*, and finally emerge as a full-grown adult honey bee. You will learn more about each of these stages of development of the castes later in the chapter.

While all three castes are all anatomically a bit different, they do share some similarities. A queen, worker, and drone bee's body each is divided into three main segments: the head, the thorax, and the abdomen.

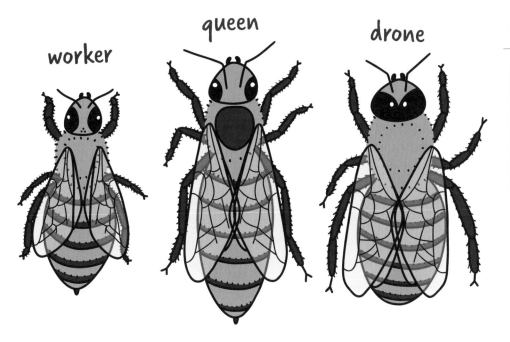

worker queen drone

A HONEY BEE'S HEAD

The *head* contains the eyes and mouthparts, along with two antennae. The antennae act as the bee's sensory system, with receptors for touch, taste, and smell. They also collect information about temperature, humidity, carbon dioxide levels, and even wind speed. Honey bees have five eyes: two large compound eyes and three simple eyes, or ocelli. The compound eyes detect color and movement, while the simple eyes detect light.

Honey bees also have a complicated set of mouthparts. These include the *mandibles* and the *proboscis*. The mandibles, made of two jaws that swing in and out from the bee's head, are used by worker bees for all kinds of household functions, like building comb and feeding the developing honey bees. The proboscis, a tube-like structure that acts like a straw and works as the bee's tongue, can switch between lapping and suction functions depending on the viscosity of the liquid the bee is drinking.

A HONEY BEE'S THORAX

The *thorax* is the area responsible for locomotion. Two pairs of wings, for a total of four wings, and six legs are attached here.

A HONEY BEE'S ABDOMEN

The *abdomen* houses the primary organs for digestion and stinging. Worker honey bees have two stomachs: one for digestion and a second, called a honey

HONEY BEE ANATOMY

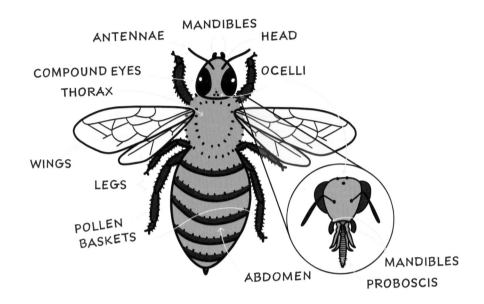

crop, which is used to carry nectar back to the colony to share with her bee family. A worker bee's stinger is the organ responsible for injecting venom into predators. The stinger is encased in a sheath, so it remains hidden and slides out only once the bee is ready to sting. Only worker bees and queens have stingers.

Each of the three castes of bees—the queen bee, the worker bee, and the drone bee—plays a vital role in the colony. As a beekeeper, it's important to understand the role of each and to be able to identify them in a colony.

The Worker Bee

Aptly named, worker bees perform all the duties required to keep the hive clean, warm or cool, stocked with food, and protected. A large managed honey bee colony can have upward of sixty thousand worker bees. Worker bees are all female and usually make up more than 90 percent of a colony's population at any given time.

worker

To aid in defending the hive, worker bees have a barbed stinger found on the end of their abdomen. Once a worker bee stings an invading animal, such as a curious raccoon, bear, or even an unlucky beekeeper, the stinger and venom sack detach from the abdomen of the bee and stay inside the victim, continuing to pump venom after the bee flies away.

She Works Hard for the Honey

Age polyethism is the division of specialized labor based on the age of a social insect. In honey bees, it describes worker bees' practice of doing different jobs in the colony as they grow older. Not every bee will hold every job, and different texts and sources vary slightly in what they report a worker bee does at different ages. However, there is general agreement about the following.

Very young worker bees are known as *nurse bees* and are responsible for caring for the queen bee in a colony and the developing bees that have yet to reach adulthood. The middle of a worker bee's life is spent on various house duties: defending and protecting, making honey and beeswax, and even removing sick and dying bees from the hive. The final third of a worker bee's life is spent as a *forager bee* or *scout bee*—taking trips to gather pollen, nectar, water, and tree resins and to hunt for a new home for the colony as necessary.

Here's how a worker bee's role in the colony will develop as she ages:

Days 1–3: The first few days of a worker bee's life are spent keeping the brood warm and cleaning cells to allow the queen to lay another generation of eggs.

Days 3–18: This is when the worker bees start to divert into different job functions, like serving as:

- Nurse bees, who are quite busy, as their sister larvae must be fed hundreds of times per day! Nurse bees also work to keep the brood nest warm, around 95°F.
- Undertaker bees, who remove any dead bees that perish in the hive. They also remove any diseased or dead brood.
- Queen's attendants, or her retinue, who feed her, clean her, and even carry out her waste.
- Heating and cooling specialists, who control not only the temperature of the hive but also its humidity. They do this by fanning their wings next to water stored in the wax comb to evaporate it.
- Architects of the hive, worker bees responsible for building the beeswax comb. Eight glands on the underside of a worker bee's abdomen are responsible for producing this wax. These glands are most productive when the worker bee is twelve to eighteen days old. Within a few days the glands begin to degenerate.

Days 18–21: Before the bees move on to become foragers, they become bouncers! These bees are responsible for guarding and protecting the hive.

Days 22–end of life: The last part of a worker bee's life is the first time she will leave the hive. Her days are spent as a forager, collecting water, propolis, nectar, and pollen. She also may be a scout bee, in charge of finding a new home when it's time for the colony to reproduce.

This is certainly painful for the predator, but a far more deadly fate awaits the worker bee. As she flies away, her barbed stinger stays lodged in the victim, which causes her gut to rip open, leading to her death shortly after she stings. During the process, a *pheromone*, an odor that communicates a message, is released to warn her sisters of the imminent danger. This self-sacrifice is one of the many examples of how worker bees place the needs of the colony above their own. She gives up her life not only to defend her colony but also to warn it of danger.

Unless called to defend her colony, a worker bee's life span is mostly related to the functionality of her wings. During the warmer spring and summer months, when older worker bees are taking many foraging flights each day, they may live only six to eight weeks. However, during the colder months, when flights aren't as frequent, honey bees may live up to five months. The more flights she makes, the more wear and tear her wings take. Once her wings are so damaged that she cannot fly back to the colony, she will die quickly without the support of her colony to help keep her fed.

The Drones

drone

Drones make up anywhere from zero to 10 percent of a colony's population and are the only male bees in a colony. They are best identified by two large eyes that meet at the top of the head—eyes that are noticeably larger than the eyes of queens or worker bees. They also have a shorter abdomen than queens and a rounder, wider body shape than either queens or workers. A drone's singular function is to mate with a queen bee—but not *their own* mother queen! Instead, drones leave the hive each day in search of virgin queens to spread the genetics of their colony. They die shortly after mating, having fulfilled their purpose.

Unlike queens and worker bees, drones do not have a stinger and cannot help defend the colony. Instead, drones rely on workers for all their needs and do not contribute resources to their own colony. Therefore, during *dearths*—periods when there is little-to-no food available for honey bees, such as winter—drones are often kicked out of the colony to conserve resources. The life span of a drone is roughly eight weeks, unless a drone has the good fortune to mate with a queen, thus ending their lives instantly.

queen

The Queen Bee

The queen is the largest bee in a colony and is the only female with fully developed ovaries, which means she is the only fertile, egg-laying bee in the colony. The queen is the mother of all the other bees in a colony, and her only job is to lay eggs. Her existence is critical. A queen bee can lay as many as 2,500 eggs each day. This is necessary to enable the colony to replace the constant turnover of worker bees, who perform all the work in a hive to keep it safe, clean, and fed— but who only live about six to eight weeks. Because the queen is the only bee that can reproduce, a colony cannot survive without her. With few exceptions, every colony has only one queen.

A queen bee has a longer life span than any of her offspring and may live up to three years. Her life span is dictated by her ability to lay fertilized eggs. If a queen runs out of the sperm used to fertilize eggs, the colony will begin to fail as the current generation of worker bees perishes, leaving no one left to care for the colony. To prevent this outcome, once a queen begins to run out of sperm, her worker bees will plot to kill her and will begin rearing a new queen bee to replace her. Therefore, a queen that is "well-mated," or has sperm from many drones in her abdomen, will have a longer life span than those that are "poorly mated," those with fewer sperm to fertilize eggs.

Queen bees do have stingers, which are used to sting other queen bees. You will learn more about how a colony rears queen bees later, but now is a good time to share that worker bees are excellent planners and will almost always

Lessons from a Beekeeper

Honey bee colonies can exhibit different personality traits. These traits can show up not only in the colony's temperament but also in the practices the colony employs. For example, some colonies are very straight beeswax comb builders. Others will go out of their way to build outside of the frames and bars you give them in your hive, sort of like coloring outside the lines.

A student of mine and I once spent some time observing the undertaker bees and saw yet another way that bees can have different personalities. The undertakers have always been my favorite of the worker bees, maybe because my mother has worked in the funeral industry for decades and passed along her good sense of humor about death. One of my first bee-keeping students, Diane, had her hives sited on gravel, so it was easy to see what debris and remains were being carried out and where they were being deposited. My hives had always rested on grass, which makes it more challenging to identify these activities. While the undertakers in one of Diane's hives disposed of their dead rather carelessly by dropping the dead bees off the edge of the hive's entrance, we noticed that the hive next door took a much more methodical approach. They carried their dead a few feet away from the hive to a sort of makeshift cemetery. They were also lining up the dead bees with the heads facing the same direction, like a burial ritual. It was too precise to be accidental and is just one more way that colonies can demonstrate their unique personalities.

produce more than one queen bee when a replacement is needed—just in case. This is beneficial because it means they have spare queens if some don't make it through the development cycle, but it also means that once the queens start to emerge, there will be some territorial fighting. In fact, the first queen that emerges will run around the hive, stinging and killing her not-yet-emerged sister queens. And unlike worker bees, queen bees do not die after stinging.

Queens are completely reliant on their daughters as caregivers. A small entourage of worker bees, called a *retinue*, is responsible for cleaning, feeding, and tending to the queen bee's every need.

The Mile High Club: Bee Sex

Bee sex is one of the most fascinating aspects of honey bee biology. A virgin queen will leave her hive five to seven days after emerging from her cell in search of some strapping fellas. Similarly, once a drone is sexually mature, about two weeks after his own emergence, his life is spent making daily flights to find a lady. *Drone congregation areas* (DCAs) are where these sexually mature drones gather to wait for virgin queens to arrive for mating. DCAs are one hundred to two hundred meters wide and five to forty meters above the ground, and they may contain several hundred to several thousand anxiously waiting drones. However, only 0.5 percent of these drones will get the opportunity to mate with a queen.

Once a queen flies to the DCA, the drones will compete to mate with her, using their superior eyesight and their speed to chase after the queen. Once a drone reaches the queen, he mounts her from behind, inserting his *endophallus*— think of this as a bee penis—into the end of her abdomen. This romantic encounter lasts mere seconds. After ejaculation, he flips backward off the queen, but his endophallus stays inserted in her abdomen, similar to how a worker bee's stinger stays attached as she flies away. And, like a worker bee that dies after stinging a victim, the drone's endophallus rips through his abdomen, resulting in his untimely death. Beekeepers call these drones popped drones for good reason, as the act produces an audible popping noise.

The bulb on the tip of the endophallus remains inside the queen, reflecting UV light, which researchers believe allows other drones to find her and possibly also prevents semen from dripping out of her abdomen as she flies. When the next drone prepares to mount her, he will remove the remains of the last drone's endophallus before inserting his own. This practice will occur again and again over the course of a few mating flights and, often, a few days. It's critical for her colony (and her own survival!) that a queen has plenty of sperm in her abdomen for many months of egg laying. Though data in published studies vary, some scientists report that queens may mate with more than forty drones, with the average

"It Wasn't Me." —Drone

Drones often get a bad rap from beekeepers. Because drones don't perform work in the hive, some perceive them as lazy and of no value. Some beekeepers also interpret the presence of drones as a sign that something is amiss in the hive. Too many beekeepers spend too much time worrying about drones: the presence of drones is not necessarily a cause for concern and usually is a sign of a healthy, strong colony! Because drones' only function is to spread the genetics of the colony, only the strongest colonies can afford to spare the resources necessary to produce and care for drones. Rather than worrying when they see a few drones in their colony, beekeepers would be better served to save that energy to invest in better understanding what the presence or absence of drones means in the hive and to the overall health of a colony.

queen mating with twelve drones.[3] Mating with drones from many different colonies helps broaden her offspring's genetic pool and can help strengthen the colony's ability to fight disease. Remember that after these early-life mating flights, queens will never mate again. Once a queen begins to run out of sperm, her workers will work to replace her with one of her own daughters.

The "Birth" of a Honey Bee

Honey bees have a four-stage development cycle: egg, larva, pupa, and adult. The development cycle for all three castes—the queen, drones, and workers—is the same, but the amount of time they spend in each stage varies between castes. The time needed to go from egg to adult bee is twenty-one days for worker bees, sixteen days for queens, and twenty-four days for drones.

For a worker bee, the cycle begins when the queen lays a tiny egg in the back of a beeswax cell. This egg has the same shape as a grain of rice. It takes approximately 3½ days for this egg to hatch into a tiny larva. Upon hatching, the larva swims around in her cell in protein-rich secretions produced by the worker nurse bees called *royal jelly*. The larva is fed this royal jelly over one hundred times per day by her sister nurse bees.

Over the course of the next five to six days, the larva is fed different combinations of nectar, honey, and pollen. On day nine or ten, the bees place a wax capping over the larva. Behind this capping, the larva will stretch out long in her cell, spin a cocoon, and develop into a pupa and finally an adult. On day twenty-one, the new adult bee will chew through the wax capping of her cell, emerging on this, her birthday! This cycle is mostly the same for queens and drones, except for the amount of time that each spends in the larval stage and pupating. One other difference is key: unlike workers and drones, which get a mix of food types, developing queens are fed exclusively royal jelly.

A queen bee can lay two types of eggs: fertilized and unfertilized eggs. Fertilized eggs will develop into female worker bees. Unfertilized eggs will develop into drone bees. This means that drones have only one set of chromosomes, with all coming from their queen mother! Workers, on the other hand, have two sets of chromosomes. The queen's ability to produce offspring from an unfertilized egg is a form of asexual reproduction.

Though the queen has the ability to lay either a fertilized or unfertilized egg, the decision of which sex to produce is not all hers. That authority lies with her worker bees. As you have learned, the drone bees' only responsibility is to spread the genetics of the colony through mating, leaving the hive each morning in search of virgin queens and if unsuccessful, returning in the evening to be cared for by their hardworking sisters. Drones consume twice as much food as worker bees and do not contribute any resources or housekeeping value to their own colonies. Therefore, not all colonies can afford to raise and care for drones at all times. Because of this, the worker bees must decide whether the colony has the resources to rear and raise drones and, when the time is right, evict the drones from the hive. The decision is based on the time of year, the strength of the colony, and the colony's access to food resources.

So how do the workers tell the queen what type of bee egg to lay? They communicate through their architecture skills: the queen bee knows what type of egg to lay based on the size of the "home" created by the worker bees to house the developing bees. You will learn more about the role of the beeswax comb in a later chapter, but as the queen's attendants guide her along the comb, she uses her feet (in the dark, moreover!) to determine the size of each beeswax cell. A larger hexagon (roughly 6.25 mm) tells her to lay an unfertilized egg, which will produce a male drone bee, and a smaller cell (roughly 5 mm) tells her to lay a fertilized egg, which will produce a female worker bee.

We now know how workers and drones are produced, but how does a colony produce a queen bee?! Queen production—the hows, whys, and whens—is a complicated facet of beekeeping. And because of that, we dedicate a whole chapter in this book to queens. But here's a quick preview: almost always, a colony will have only one active queen bee. Queens are very territorial and will fight to the death to ensure only one queen mother remains.[4] However, queens, like

THE FOUR
Honey Bee

Days	1	2	3	4	5	6	7	8	9	10	11	12
	EGG			LARVA								

WORKER

DRONE

QUEEN

STAGES OF
Development

13 14 15 16 17 18 19 20 21 22 23 24

PUPA

ADULT BEE!

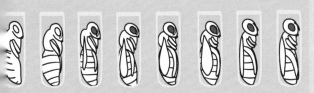

any organism, will not live forever and do have a life span, albeit one many times longer than their offspring. Although queens can choose to lay worker eggs or drone eggs, "queen eggs" don't exist.

Worker bees create queen bees from worker bee larvae. You'll learn more about this process in chapter 6. Remember that queen larvae, unlike worker and drone larvae, are fed exclusively royal jelly. Once a worker larva hatches from an egg, nurse bees will feed this new larva exclusively royal jelly, and more of this brood food than normally fed to worker bee larvae. This simple diet change produces a modified version of a worker bee that has the ability to biologically reproduce: a queen bee.

Colony Reproduction

We've discussed how reproduction works in a colony: a queen bee is responsible for laying all of the eggs in a colony. However, remember that a honey bee colony works as a unit called a superorganism, and this superorganism also needs to be able to reproduce to produce new colonies of honey bees. This phenomenon is known as *swarming*. A colony that casts a swarm has the potential to make two colonies out of one.

The mechanics of a swarm are fascinating: when ample food resources and a robust, healthy queen results in a strong colony, it signals to the workers that the colony can reproduce. That is, it can divide into two or more colonies. Also, if the queen starts to run out of space to lay eggs, either naturally or because of misguided decision-making and interventions by the beekeeper, this also can signal the colony to swarm.

When a colony swarms, roughly half the colony will leave with the original queen bee to find a new home. The bees that are left behind will begin to rear new queen bees before the swarm leaves so that a few days after the swarm departs a new queen can emerge, perform her own mating flights, and return to begin laying eggs to ensure the original colony has a laying queen. If all goes as planned, two colonies with two laying queens will now exist. You will learn more about swarming, and how to work to prevent it in your colonies, in chapter 8.

Honey Bee Communication

Honey bees have advanced communication abilities. Their primary communication tools are their two antennae and pheromones.

The antennae of a honey bee are divided into three parts. The two segments nearest the bee head are responsible for movement of the antenna. The section of the antennae farthest away from the head is called the flagellum, which is further divided into smaller subsegments. The flagellum even indicates the sex of the bee. In most species of bees, including honey bees, females have ten subsegments and males have eleven.

These antennae are a powerhouse of communication, as different sensors and receptors can identify surface texture, humidity, carbon dioxide, gravity, shapes, temperature, and pheromones. They can even perceive sugar concentrations. They also help the bee determine its flight speed and can detect the vibrations of other bees' movements. You will learn later how honey bees perform a dance to communicate where to find food resources. The antennae are critical for this, because they help the bees interpret the dance's vibrations even in the total darkness of the hive.

Pheromones are secretions that send messages to other organisms, and they are intended to elicit some response. A colony has many different types of pheromones, and these pheromones are grouped into two types: releaser pheromones and primer pheromones. Most of the pheromones in a colony are releaser pheromones, which means they initiate an immediate behavioral response in the bees in a colony. The alarm pheromone you learned about earlier is one type of releaser pheromones. On the other hand, a primer pheromone acts at the physiological level, triggering complex responses that can result in behavioral and developmental changes. Pheromone capabilities can vary by caste and can even change over time. For example, only worker bees can produce alarm pheromones, and the pheromone is weak in young bees, increases a few weeks into a worker bee's life, and then decreases again as she moves to the foraging function.

A few of the more notable pheromones in a colony include queen mandibular pheromone, alarm pheromone, Nasanov pheromone, footprint pheromones, and brood pheromones.

QUEEN MANDIBULAR PHEROMONE

Queen bees have a number of pheromones produced by different glands that collectively are known as the "queen signal," but the most studied of these pheromones is the *queen mandibular pheromone* (QMP), first discovered in 1960. The QMP is what attracts the eight to ten workers that make up the queen's retinue and spurs them to recognize her and care for, clean, and feed her. The QMP also causes a swarm to cluster around the queen bee, protecting her during this process. The QMP regulates and stimulates worker bee activities, ensuring they are continuing the necessary activities to keep the colony clean and fed. It's also this pheromone that attracts drone bees for mating. If this pheromone starts to fade, it signals to the worker bees to begin the steps to rear a new queen to replace her. Interestingly, beekeepers are able to manipulate a colony's inclination to rear a new queen when her pheromones are absent—which signals that the colony is queenless—to rear queen bees! Because the QMP helps a colony identify its mother, if a beekeeper needs to introduce a new queen into a colony, they must introduce the queen slowly to allow the bees to become familiar with her odor first. Otherwise, the colony will mistake the new queen for an intruder and kill her.

ALARM PHEROMONE

Pheromones are also critical in a colony's defense system. After a worker bee stings a predator, she flies away, leaving her stinger behind in the skin of the unfortunate animal or human. In the process, her stinger is ripped out, effectively disemboweling her and releasing a pheromone that warns her sisters of danger nearby. This is how a careless intruder may suffer numerous stings rapidly if a honey bee colony is disturbed. Further, the defender bees at the entrance of a hive can release this pheromone if they feel the colony is being attacked. Fun fact: this pheromone is the same chemical used to produce artificial banana flavoring in candy and other foods! Have you ever heard that beekeepers should avoid eating bananas right before checking their hives? The idea is that the banana scent could alert the defender bees, making your hive inspection that much more tricky.[5]

NASANOV PHEROMONE

The Nasanov gland is a tiny gland toward the end of a worker bee's abdomen that releases another releaser pheromone used to help "mark" sites for her sister worker bees. A worker honey bee releases this pheromone by flexing the tip of her abdomen downward, while elevating her abdomen and fanning her wings to help disseminate the pheromone. This marking pheromone can be used to mark the hive entrance, assist with clustering during swarming, and mark locations of food sources. Beekeepers often see this phenomenon when opening a hive and disturbing the brood nest. Worker bees, sensing that some bees may have lost their way during the inspection, will stand at the entrance of the hive, releasing and fanning this pheromone to help their sisters find their way back home.

FOOTPRINT PHEROMONES

Footprint pheromones are present in all three castes, and they are located at the end of the legs in the tarsal glands. Similar to the pheromone emitted by the Nasanov gland, the tarsal gland emits a pheromone used for marking. In worker bees, these footprint pheromones allow a worker bee to mark the hive entrance and food sources. The footprint differs from the Nasanov pheromone because it is active at short distances, whereas the Nasanov pheromone can be cast at longer distances. In queen bees, this pheromone inhibits worker bees from rearing new queen bees. Though they know that drone bees also have tarsal glands that emit a different pheromone from that of the other castes in a colony, scientists are still unsure of the purpose of this pheromone in drone bees.

BROOD PHEROMONES

Salivary glands in the larvae secrete ten different compounds that make up brood pheromones. These compounds can vary based on the age of the larvae. For example, pheromones from older larvae have been found to induce worker bees to cap the cell with beeswax. These pheromones also increase the activity of the workers' *hypopharyngeal glands* to produce the royal jelly that is fed to young larvae. Brood pheromones also regulate pollen collection and how quickly worker bees progress to the foraging function.

Both QMP and brood pheromones have another important role in the colony: suppressing worker bee sexual development. An absence of QMP and brood pheromones for an extended period of time can create quite an interesting set of circumstances. Workers do not lay eggs and, therefore, they have no need for ovaries. However, worker bees do actually have ovaries that are suppressed by these pheromones. If a colony goes without a queen, eggs, and larvae for many weeks, the lack of pheromones will cause the workers to develop ovaries, and then they can actually begin to lay eggs.

More eggs in a colony may sound like a positive development. However, though workers can develop ovaries, only the queen bee has the ability to mate and carry sperm to produce fertilized eggs. Because the worker bees are not mated, they can only lay unfertilized drone eggs—that is, they can only produce a hive full of males. The reason for this fascinating phenomenon is genetic preservation. If a colony loses its queen and is unable to produce a new one, or if the new queen fails to return to the hive after her mating flight, the colony will perish once the original queen's last eggs hatch, develop into adult worker bees, and die. Therefore, the workers lay drone eggs as a last-ditch effort to spread the genetics of the dying colony by producing males that can mate with queens from other hives. This is what's known as a laying-worker hive, and it's quite a pain for the beekeeper to try to correct the problem and save the colony! But it can be done, and you'll learn how in chapter 6.

CHAPTER 2

A Honey Bee's Home

You will learn later how to select a hive as a home for your bees, but first it's important to understand that the human-made hives we provide to our bees are simply a protective shelter for the home the bees build for themselves—the comb made out of beeswax. In the wild, honey bees will also look for something to protect their beeswax home. They will most often choose to build their beeswax comb in a covered area, such as inside a tree trunk or cave, or even in less natural areas such as an owl box, an empty RV, a storage shed, or the walls of a house.[1] Remember that beeswax is produced from glands on the abdomens of worker bees that are twelve to eighteen days old.

Inside their protected vessel, whether it's a wooden hive built by a beekeeper or a tree trunk, the worker bees will make rows of thin sheets of connected hexagons made of beeswax. These sheets of beeswax have hexagonal cells on both sides and are what we call the beeswax comb, or *drawn comb*. These cells are where all the bee magic happens:

- These hexagons are where the queen will lay her eggs, one generation after the next. She will lay one egg in each cell, and within these cell walls the egg will hatch into a larva, spin a cocoon to develop into a pupa, and eventually emerge as an adult bee.

PLACES I'VE FOUND BEES

a storage shed

a compost bin

an owl box

a toilet

an empty RV

- The beeswax hexagons are also used as a storage space, similar to a pantry, for the food that feeds the colony: pollen that will ultimately be fermented into bee bread for the developing larvae, nectar that will eventually be dehydrated into honey for the adult bees to eat, and water.
- Water stored in the beeswax comb can even help to cool the hive, as the bees will fan their wings to create a sort of evaporative cooling process during the hot summer months.
- Finally, the honey bees use the beeswax comb as a tool for communicating messages. Because most colonies exist in total darkness, workers can't see their sisters, so they rely instead on feeling vibration messages sent as worker bees move across the comb.

Beeswax Comb

To make beeswax, the worker bees must consume the carbohydrates in nectar and honey, which is then metabolized to produce thin scales of wax secreted from eight small glands located under their abdomens.

Bees must eat upward of *eight pounds of honey* to produce one pound of beeswax! That's eight of the bear-shaped bottles of honey you see at the grocery store just to make one pound of beeswax comb. Because so much honey is required to produce beeswax, honey bees can produce beeswax only during times of year when there is an abundance of nectar-producing plants providing a continuous stream of carbohydrates for energy. For most places around the world, that means bees can produce beeswax only during a very short time of the year. This period also coincides with the time when your bees will be able to make honey. Later, when we discuss honey bee nutrition, you will learn that you also can feed your new colony a mixture of white sugar and water to provide the carbohydrates needed to build out the beeswax comb that makes up their home if they aren't able to gather enough nectar on their own.

Once the beeswax is secreted from the glands on their abdomens, the worker bees will move the wax to their mandibles, or jaws, and form the beeswax into hexagons. These hexagons are joined to form large sheets of beeswax comb, with hexagons on both sides of the comb. These sheets are always built perpendicular to the ground for strength and stability. Honey bees even build these cells in a way to help prevent "spillage" out of the cells: the sides of the cells are tilted upward at an angle of roughly 9° to 13°.

Beeswax hexagons are critical to a colony, as without them the queen would have nowhere to lay her eggs and the bees wouldn't have any place to store pollen, bee bread, or nectar, or to make honey.

But Why Hexagons?

The hexagonal shape bees create from beeswax to build their home provides several advantages. First, hexagons are a model of efficiency. Hexagons are one of only three regular shapes that can provide tessellation, which refers to the covering of a surface with a repeating pattern that has no overlaps or gaps. (The other two shapes are the equilateral triangle and the square.) Of these three, hexagons provide the most storage space per inch while using the least amount of building material. Furthermore, hexagons can cover a curved surface with minimal material waste. And though the goal of bee-keepers is to provide a space for our bees that encourages them to build straight comb, this is simply because it's easier for us to then remove the sheets of comb to inspect the hives. Left to their own devices, honey bees will use the space as efficiently as possible, and in the wild that often means sheets of comb that twist and curve. Charles Darwin, in his book *On the Origin of Species*, writes that the comb of honey bees is "abso-lutely perfect in economizing labour and wax." Hexagons are also a shape that provides great strength. The open lattice pattern provides a very strong vessel without being too heavy, which is why hexagons are used frequently in architecture.

A Bee's Nest in the Wild

It's important to understand that honey bees are not a domesticated species, and the behaviors of wild honey bees are really not very different from that of managed honey bee colonies. The primary difference is that managed honey bees are kept in hives that can be manipulated and inspected, and managed colonies are similar to livestock in this manner. As a beekeeper, you'll install your bees in hives with removable frames or bars, encouraging your bees to draw comb in straight sheets instead of the organic patterns bees choose in the wild. These removable frames and bars allow you to manipulate and move sheets of beeswax comb around without breaking or damaging the comb. Beekeepers can rearrange frames and bars within the hive and can even share these resources with other hives in the *apiary*. These management techniques can be used to help promote the health of the colony and meet our own goals as beekeepers.

Understanding how and why a colony structures its comb and nest a certain way helps better inform decisions by the beekeeper when manipulating the

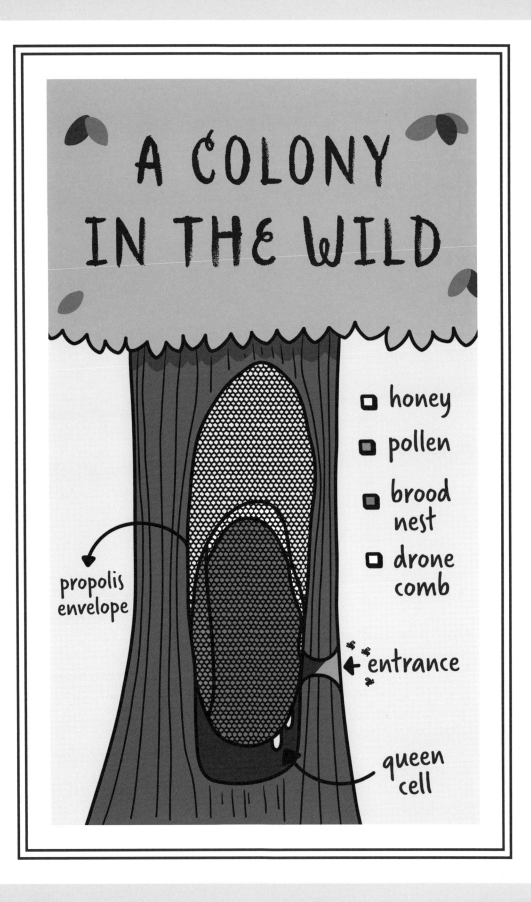

frames or bars. If these manipulations are performed without the knowledge and understanding of how they may disrupt the natural activities of the colony, they can cause detrimental consequences to the colony. We will cover what these manipulations are and how to perform them later, but before we do, let's build an understanding of the makeup of a natural honey bee nest.

Wild honey bee colonies will generally choose protected, darkened enclosures with a small entrance of just a few inches in length or diameter to house the colony. This is because a small entrance means less space to defend. In the wild colony, the queen will lay her eggs in a compact cyclical pattern, which will eventually produce a bullet-shaped brood nest. If a colony is strong enough to rear drones, the workers will place those drone cells around the outer edge of the brood chamber.

The reason for this is fascinating: drone brood is, in several ways, the hive's sacrificial lamb. First, during the early spring, a hive may experience warmer days but still have cooler nights. When the temperature is below roughly 55°F, the hive will begin to cluster tightly around the queen. Any brood that falls outside of that warm cluster will perish. A colony is not significantly affected if drone brood die because they will not contribute to a colony's immediate needs or activities. However, any dead worker brood can prevent the colony from growing its workforce. As a result, drone brood around the outer edge of the brood nest ensures that any warmth generated is used to keep worker brood, the most important brood, warm. Additionally, the drone brood can help provide warmth to the worker brood in the center of the brood nest. This perimeter of drone brood also can provide some level of protection against predators. If a predator, such as a raccoon or skunk, were to break into the brood nest, the animal may only get away with a handful of drone brood before being chased away by the defender bees.

The colony will also store pollen, honey, and nectar around the brood nest. By placing food around the periphery of the brood nest, worker bees have quick and easy access to pollen, bee bread, and nectar to feed the developing larvae. Honey is also a good insulator and can help regulate the temperature of the colony's brood nest.

In nature, honey bees may or may not build straight comb, but honey bees, whether managed or wild, follow what is called the rule of bee space. *Bee space* was first identified in 1851 by Lorenzo Langstroth, who would later be known as the inventor of the modern-day *Langstroth hive* design. He discovered that bees will fill any space less than ¼ inch with propolis, and they will build comb in any space ⅜ inch or larger. Building in this manner allows bees to be efficient and ensure no space goes unused in the hive.

Lessons from a Beekeeper

My first lesson in listening to the needs of my bees began before I had honey bees! Though I am not a handy person, I built my first hive with friends following plans we found on the internet. (To this day I still wonder how such a clumsy person so lacking in the ways of carpentry, tools, and general tradesman skills can be such an accomplished beekeeper. I truly believe my abilities to listen to and observe my bees are the only reason I can overcome such ineptitude.) My first hive was *not* a work of fine carpentry, though the sides mostly met where they should. I remember an overwhelming feeling of pride that I turned a pile of lumber and some internet architecture plans into a new home for my bees. In planning the design, I read somewhere that creating a screened bottom for the hive, as opposed to one made of solid wood, was important to help the colony with ventilation and would help the bees stay cool. I didn't yet understand the needs of bees or their incredible ability to self-regulate the temperature of their hive, but the logic made sense to me. It's hot in Texas! If I lived in a wooden box with no air conditioning, I'd be desperate for some air too! I affixed a window screen instead of a solid board as the bottom piece of my hive.

Within a few months of installing my colony of bees, my bees used propolis to seal up every inch of that window screen! What was happening? Why did they hate the home I gave them!? Later, I would learn more about the wants and needs of bees and their abilities: bees don't like light in their colonies, preferring to work in darkness. They also don't like excess gaps and openings, because openings have to be guarded to protect the colony from intruders. Finally, bees actually do a phenomenal job controlling not only the temperature but also the humidity levels of their hive. All my screen did was let light shine in, provide ways for intruders to enter the hive, and work against their efforts to adequately control the hive's climate. My efforts to help my bees actually created a great deal more work for them.

Nevertheless, beekeepers continue to debate the benefits of screened versus solid bottom boards. There are some valid arguments that a screened bottom board may help with mite control because as hygienic bees clean small mites off one another, the mites fall through the screen below and out of the hive. What is clear is that my bees demonstrated that a screened bottom board did not let them live their best life, and it taught me that in times of uncertainty, the bees will tell you what they need. You just have to listen.

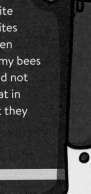

If you are interested in learning more about how bees choose their homes in the wild, check out Dr. Thomas Seeley's book *Honeybee Democracy*. Dr. Seeley's life work has been dedicated to studying swarms, and a great deal of what we know about hives in the wild is derived from his research.

Identifying What You See in Your Hive

Identifying the elements in a colony is a difficult task for new beekeepers. Without an experienced mentor on hand to point out the differences among the castes or the various stages of the development cycle, it can be intimidating to figure this out while working amid a flurry of buzzing bees. Study this section carefully and then, during those early hive checks, practice spotting and identifying all these elements. The simple act of knowing what you are looking at will progress your beekeeping skills faster than almost anything else you can do.

BEESWAX COMB

When bees initially produce beeswax, it will be stark white in color and almost translucent. As it is constructed into comb and continues to get used, the comb will grow darker because of staining from pollen oils and the propolis that bees use to line the beeswax cells before a queen lays an egg. Furthermore, with each generation of emerging brood, bits of the cocoon will be left behind in the cell, hardening the wax and darkening the comb until it grows brown and then almost black. This fairly rapid change in color can be concerning to new beekeepers, but this is completely normal! Later, we will discuss ways to recycle and replace a few of the oldest frames each year during your early spring checks.

QUEEN BEE

A queen is most easily identifiable by her long abdomen, which has a slight bullet shape. While drones also have large abdomens, their shape is wider and more rounded. Drones are often mistaken for queens because of their larger body shape. If you are unsure, check out the wings: do they extend to the end of the abdomen? If so, you may be seeing a drone! A queen's abdomen extends farther than her wings. Queens also do not have hair on their thorax, unlike their offspring, and they also move differently than the rest of the bees in a colony. It's a nuanced difference but one that beekeepers will notice the longer they study their bees. A queen's movement is a more fluid movement, and one with purpose. The workers, on the other hand, tend to dogpile over one another with movements that are more frenetic. I once read that queens move as if they are waltzing, whereas workers move as if they were in a mosh pit. Finally, queens often have less pronounced stripes on their abdomens than workers. It's not a hard rule, but I often spot my queens by the muted colors on their abdomens.

DRONE BEE

Drones' most identifiable feature, after their wider abdomens, is their eyes. A drone's much larger eyes meet in the top-center of its head and provide it with superior eyesight so that it can scan the sky for a potential mate, a queen bee zipping by in flight.

WORKER BEE

Worker bees are, by far, the most numerous caste in the hive, so you won't have any problems locating them! Workers are the smallest individuals in the colony, with a much narrower abdomen than drones and a shorter abdomen than the

Queen bee. © 2023 by
Tara Chapman.

Drone bee. © 2023 by Tara Chapman.

Worker bee. © 2023 by Tara Chapman.

queen. Workers, like all three castes, emerge as fully grown adults, so they do not change size over time. Furthermore, as workers rotate through job functions, there are no visible changes in their physicality, with one exception: newly "born" worker bees do not have fully formed *exoskeletons*, and their coloring is a bit ashy and gray. An exoskeleton refers to the external skeleton that supports and protects the honey bee body. Older worker bees are distinguishable by the little hair that remains on their bodies and their tattered wings, which will begin to show signs of wear as they become forager bees and spend their days flying to and from the hive.

EGGS

The first stage of the development cycle is the egg. An egg resembles a tiny grain of white rice that adheres to the back wall of a beeswax cell.

LARVAE

Both worker and drone larvae grow very, very fast after hatching from an egg and will be 1,500 times their original size when it's time to pupate! Newly hatched larvae are very small compared to their cell size and swim around in their cell, feeding on the royal jelly produced from the heads of the nurse bees. The description that resonates most with me is that a newly hatched larva resembles a tiny string floating in a milky substance. Within just a few days, the larva grows exponentially, filling up the cell and curling into a C-shape. Healthy larvae should be white, not yellow or discolored, and will be plump and squishy!

Eggs laid by a queen bee.
© 2023 by Tara Chapman.

Worker bee larvae.
© 2023 by Tara Chapman.

Capped worker brood.
© 2023 by Tara Chapman.

Pollen and bee bread stored in beeswax
comb. © 2023 by Tara Chapman.

PUPAE

Pupae are at the only stage of the development cycle hidden from the beekeeper's eyes: at the tail end of the larval cycle, the honey bees place a beeswax cap over the top of the cell. If you were to pull the cap off, you would see the head of the developing bee looking right back at you. Beekeepers refer to this stage of the brood cycle as *capped brood*. (Similarly, beekeepers refer to eggs and larvae as *uncapped brood*, or *open brood*, because these stages are not covered by the beeswax cap that is present over the pupal stage.) Both drones and worker bees develop in the beeswax cells, parallel to the ground with their heads facing outward. Queen bees, on the other hand, develop with their heads facing down toward the ground. These cell caps can range in color but are usually light to dark brown and are always porous to allow oxygen into the cell.

POLLEN

Different flowers produce many different shades and colors of pollen, and when arranged in the comb, pollen can present a rainbow of colors in the beeswax comb. Honey bees usually pack pollen in adjoining cells, so you should see dozens of cells filled with pollen in certain areas of the hive, often just around the perimeter of the brood nest. When foraging bees return to the hive with the pollen, they use their heads to pack it in the back of the wax cell, so look for a smooth, solid, matte surface. And while orange and yellow may be the most common pollen colors, you might be lucky enough to spot gray, blue, red, or purple.

Nectar stored in beeswax comb.
© 2023 by Tara Chapman.

Honey stored in beeswax comb.
© 2023 by Tara Chapman.

BEE BREAD

If you are seeing the same solid rainbow of colors but with a shiny surface, you have likely discovered bee bread! Once the pollen is packed tightly into the cell, the honey bees will seal the pollen with a bit of honey or nectar to preserve its benefits. This causes the pollen to ferment and gives the surface a sheen. Using a knife, carefully cut out a small section of the comb filled with bee bread. Pop the plug of pollen out of the cell and admire the many colored layers. You can even eat the bee bread! I much prefer its texture and flavor to that of plain pollen.

NECTAR

Nectar stored in the comb is in liquid form, is very shiny, and will catch your eye because it will reflect light. If you take a frame of nectar and shake the frame vigorously, the liquid nectar will easily shake out of the frame.

HONEY

Honey is nectar that has been preserved through a dehydration process and then given a beeswax cap. The caps should not be porous and are usually translucent so you can see the honey underneath the cap. (As the comb in your colony ages and grows much darker with use, however, the honey stored in old brood comb can have a much darker appearance that can be confusing to beekeepers. This darkened surface can look similar to the cap that is placed over pupae. If you're unsure, look closely at the surface: capped pupae will always have a porous surface, unlike honey.) If you take a frame of honey and shake vigorously, it will not shake out of the comb like it would if it were nectar!

Remember, the size of the cell communicates to the queen whether she should lay a fertilized egg to produce a worker bee or an unfertilized egg to produce a drone bee. Therefore, we can identify workers from drone cells by the size of the beeswax cell. Worker cells are noticeably smaller than drone cells, around 5 mm wide, whereas drone cells are 6.25 mm. Another difference is that the drone pupae, once capped, will protrude noticeably outward from the beeswax comb, while worker bee capped brood lies flatter to the plane of the comb. Whenever I see capped drone brood, it always makes me think of the corn pop cereal I had as a child.

Queen cells, which will house developing queens, will only be present in your hive if the colony senses your queen is failing or has failed. They will also rear new queens if they are preparing to swarm. This book has an entire chapter dedicated to queen bees, and you will learn more about the different types of queen cells then. For now, know that queen cells sit on top of the comb, not within it like the others, and are large and elongated. They can be found in the middle or at the bottom of a frame and resemble peanuts, particularly once they are capped by the worker bees. But don't be fooled by nubbins. "Nubbins" is the term beekeepers use to refer to "practice" queen cells. These are the start of a queen cell and have fooled many a beekeeper. If you see what you believe to be a queen cell and it is not capped, hold the frame so that you can peek inside the cell. If you don't see an egg or a developing larva, you have found a nubbin! These may be used later, should the colony determine the need to rear a new queen.

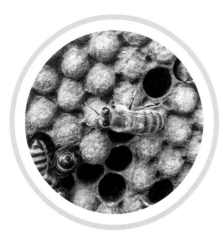

Capped drone brood.
© 2023 by Tara Chapman.

Queen cell containing developing
queen bee. © 2023 by Tara Chapman.

CHAPTER 3

Honey Bee Nutrition

Before we get into when, how, and why you might need to feed your bees, successful beekeepers must first understand the basics of honey bee nutrition and the needs of each caste. Generally, I find that few first-year beekeepers understand this topic, but it is easily one of the most important. As with humans, poor nutrition can result in devastating health effects on the colony. However, while human lives are usually measured in decades and a few weeks of poor nutrition likely won't have any long-term negative effects, the life span of honey bees is measured in weeks. A few weeks of poor nutrition in the developmental stage can wreak havoc on the colony when those same bees are expected to become the foraging workforce to gather food for the rest of the colony. If the timing is right, and this weakened workforce comes of age during those few precious weeks when bees are expected to make honey, the colony might forfeit its ability to store enough honey for the winter and will starve to death. On the other hand, too many beekeepers with a poor understanding of the needs of bees often treat them as pets and believe they require daily feeding, stuffing the hive full of pollen substitutes and sugar water. Though these beekeepers are well intentioned, their colonies can experience grave consequences as well.

Understanding the nutritional needs of your bees and knowing how and when they can access these nutrients from their environment will help you know when feeding is necessary.

The Nutrients Required by Honey Bees

Honey bees require three different *macronutrients* to maintain colony systems, energy, and growth: carbohydrates, protein, and lipids (fats). Macronutrients are nutrients derived from food sources that make up the largest portion of an organism's diet. Let's review each of these macronutrients, their importance in a honey bee colony, and the food sources that provide them. If you've done any study of human nutrition or tracked your own macronutrient intake, this will all sound very familiar!

Carbohydrates: Carbohydrates provide fuel and energy for the colony to function. The primary sources of carbohydrates for the colony are both nectar and honey.

Protein: Protein and amino acids are necessary for muscle and glandular development and for repairing tissues. The primary source of protein and amino acids for a colony is pollen.

Lipids (fats): Lipids include different fats and essential fatty acids such as omega-3 and omega-6. Lipids are necessary for brood rearing, and a proper balance of omega-3 to omega-6 fatty acids is necessary for good brain health. Honey bees obtain lipids from pollen.

Additionally, honey bees require a number of *micronutrients*: vitamins and minerals that are necessary in much smaller quantities than macronutrients. Vitamins are obtained from pollen, while minerals are obtained from pollen, some nectars, and from standing water. Nectars that result in darker-colored honey often provide a higher concentration of minerals.

What's the Deal with Omega-3 and Omega-6 Fatty Acids?

Much study has been done on the ratio of omega-3 to omega-6 fatty acids in both humans and honey bees. Scientists believe that humans evolved on a diet with a ratio of 1:1 of these two fatty acids. However, the modern Western diet has a ratio closer to 16:1 of omega-6 to omega-3 fatty acids. This modern diet, deficient in omega-3 and with excessive amounts of omega-6, has been found to promote pathogens and illness such as cardiovascular disease, cancer, and inflammatory and autoimmune diseases.[*] In the case of honey bees, studies have found that a high omega-6 to omega-3 ratio leads to reduced brood rearing and increased mortality rate.[†] This diet also resulted in smaller hypopharyngeal glands, which is the set of glands along the side of worker bees' heads that produces royal jelly. Royal jelly is fed to all young larvae and queen bees, so smaller glands means less of this critical resource. Additionally, overdevelopment and the use of land to grow monoculture—a farming initiative where a single crop is grown across large swaths of land—have contributed to poor nutrition for honey and native bees. If you're suddenly inspired to include more omega-3 in your own diet, look to avocados, peanuts, flaxseeds, and some seafoods such as sardines, salmon, and oysters.

* Artemis P. Simopoulos, "The Importance of the Omega-6/omega-3 Fatty Acid Ratio in Cardiovascular Disease and Other Chronic Diseases," *Experimental Biology and Medicine* 233, no. 6 (June 2008): 674–688.
† Yael Arien et al., "Effect of Diet Lipids and Omega-6:3 Ratio on Honey Bee Brood Development, Adult Survival and Body Composition," *Journal of Insect Physiology* 124 (July 2020).

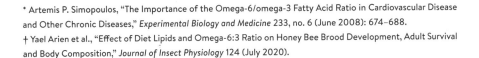

The Hunt for Nutrition

Now that you understand the different macro- and micronutrients in a colony and how they are obtained, let's discuss how a colony finds and collects nectar and pollen, the two external food sources that provide the three macronutrients, and how they are processed and fed to the bees in the colony. The nutritional needs of developing bees at different stages of development are not all the same. Additionally, the nutritional needs of the different honey bee castes can vary depending on their jobs and how they are contributing to the growth and development of the colony.

UV VISION: A BEE'S SUPERPOWER

As you already know, foraging bees will leave the hive on days when the temperature is about 50 to 55°F or higher to gather pollen, nectar, water, and sap or resin for propolis. These foraging bees can fly three to five miles each way in search of food.[1] The foraging bees' brains have a complex system of mapping abilities that allows them to travel long distances and find their way back home.

Adult honey bees have many impressive traits that allow them to be very adept foragers. In addition to their impressive mapping abilities, they have a very keen ability to detect odor with their antennae. Honey bees can see every color that humans can see except for the color red, which they perceive as a sort of dark gray. But they have a unique ability that humans do not possess: they can see in the ultraviolet (UV) spectrum. As a result, plants have evolved to include different markings within the UV spectrum that are not visible to the human eye to attract honey bees. These markings and patterns found on some flowers are called nectar guides and help guide the honey bees to places to find food. This adaptation ultimately helps ensure the flower gets pollinated!

Honey bees can also see polarized light, which is very useful because it helps them navigate relative to the sun, *even if the sun isn't shining*! Honey bees also have the ability to process colors very quickly, which allows them to identify individual flowers rapidly during flight.

CAN HONEY BEES ... DANCE?!

Honey bees are what is known as forage consistent: this means that they will limit their visits to a single plant species until the food provided by those flowers is exhausted. This behavior is what makes them such great pollinators, as pollination requires the transfer of pollen between different flowers on the same plant. The fact that a bee will continually visit a plant species over and over again increases the chances that pollination will occur.

When a bee finds a profitable source of food, she will return to the hive to communicate the location of this food source through a motion called the *waggle dance*. She will move her body in a figure eight shape across the comb, and when she reaches the middle of the figure eight, she will shake her body rapidly. The angle of the dance communicates the angle of the sun to the flower. The duration of the waggle communicates distance: the longer the waggle lasts, the farther the food source is from the hive. You will learn how this dance is also used by a swarming colony in chapter 8.

Can we have a dramatic pause here and acknowledge how unbelievable this is? When's the last time you even went to the grocery store without the use of a mapping app on your phone?

"YOU MAKE ME FEEL

1

A bee moves in a figure 8 across the beeswax comb.

She will move her body rapidly—hence the waggle.

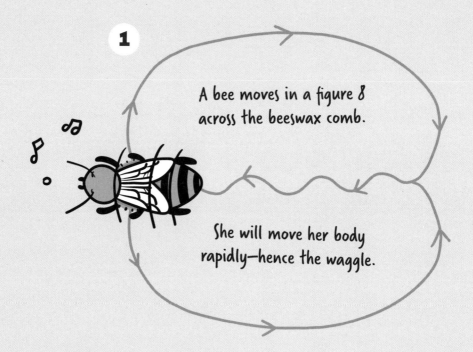

3

The duration of the waggle communicates the distance: the longer the waggle, the farther the food source from the hive.

LIKE (WAGGLE) DANCING."

2

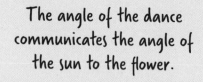

The angle of the dance communicates the angle of the sun to the flower.

PHEW!

POLLEN IN THE COLONY

Pollen provides most of the protein and amino acids for a colony. Forager bees collect pollen on special structures on their two hind legs, which beekeepers call pollen baskets. This is fun to see in a hive, as it appears that the workers are wearing brightly colored leg warmers. But those legwarmers can be heavy: forager bees can carry up to 35 percent of their body weight in pollen![2] That would be like a 150-pound person carrying 52 pounds of groceries for three to five miles! Once they return to the hive, the workers will scrape off the pollen into a cell, and other worker bees, using their heads, will pack the pollen into the cell. These workers also add glandular secretions, honey, and nectar to the pollen mix. This recipe results in the fermentation of the pollen, a substance known as bee bread. Though much is still unknown about bee bread, it is now commonly accepted that the fermentation of the bee bread is to preserve its nutritional benefits—much the same reason that humans ferment foods.

Pollen is critical to brood rearing, and without it a colony cannot create more bees. Pollen is necessary for nurse bees to create the royal jelly fed to the youngest larvae, and nurse bees also directly feed pollen to older larvae. In fact, pheromones released by the brood in a colony spur the foragers to collect more pollen!

First, young adult worker bees will consume pollen and bee bread seven to ten days after emerging from their cells. This is because their hypopharyngeal glands, which produce royal jelly, are underdeveloped when new worker bees first emerge from their beeswax cells, and only the consumption of pollen allows for their full development.

The young worker bees must also consume pollen to produce royal jelly. Royal jelly is a milky white substance made of about 60 percent water. The rest of the royal jelly is mostly proteins and sugar, with a small amount of lipids and minerals. This royal jelly is then fed to newly hatched larvae of all three castes and adult queen bees. Drone and worker bee larvae consume this royal jelly exclusively for the first three days after hatching and then are switched to a diet of bee bread made from pollen until they begin pupation. Unlike pollen and nectar, however, royal jelly cannot be stored in the hive for later use. Nurse bees feed it to larvae by immediately collecting the royal jelly from the hypopharyngeal glands on the sides of their heads and placing it in the cell with the larvae. To the beekeeper, this looks like tiny larvae swimming around in a pool of milky liquid. But while worker and drone bees go on to eat nectar and honey after they emerge as adults, queen bee larvae and adult queens exist solely on this royal jelly diet. Later, you will learn how the timing and ratios of feeding royal jelly will determine how and whether a queen bee is reared in the colony.

If pollen helps the colony produce more bees but these bees only consume pollen and bee bread in their early stages of life, what is their source of nutrition as they grow older? Carbohydrates, carbohydrates, carbohydrates! Older bees' primary nutrition source is the sugars obtained from nectar.

Foraging bees will visit a flower and draw up nectar in their proboscis. This proboscis, approximately 7 mm long (0.28 inches), has two "settings": a lapping setting and a suction setting. Interestingly, they will switch between these two modes based on the viscosity of the nectar. A lower-viscosity nectar calls for sucking, whereas a higher-viscosity nectar can only be obtained through lapping. Have you ever tried to suck up a really thick milkshake through a straw? It's pretty tough to do, maybe even impossible. Rather, using a spoon is a bit easier! It's the same with a bee's proboscis. The bee can alternate between lapping and sucking depending on how thick, or viscous, the nectar is.

Worker bees have two stomachs; one is for digesting food, like a human stomach, and one is for storage. The bee will store nectar in her second stomach, which is called a honey crop. The honey crop acts as a container to carry nectar back to the hive, and a small flap separates the honey crop from the bee's own digestive stomach. This flap can allow small amounts of nectar to pass through to help provide energy to the bee during flight. It can even grind down and filter pollen out of the nectar to keep it out of the bee's digestive tract. In fact, the processing of nectar into honey actually starts right here, in the honey crop, as enzymes and probiotic bacteria are added to the nectar. Once back at the hive, the foraging bee regurgitates the nectar from her honey crop and passes it to receiving bees, who will then deposit the nectar into the cells, also through their own regurgitation. (This is why honey is sometimes cheekily referred to as "bee vomit"!)

It's important to note that in most climates, bees do not have access to nectar year-round. Depending on your climate and growing season, your bees may only be able to make honey as little as six to eight weeks out of the year. The period that bees have access to enough nectar that they can begin to store it as honey is what's called a *honey flow*, or *nectar flow*. During a nectar flow, honey bees act as phenomenal little hoarding machines, as they have to be able to collect enough nectar to feed the colony for many months. But nectar has a finite shelf life. Despite the fact that it's often almost half sugar, the remainder is water, and water provides an ideal environment for fermentation to occur. Honey bees will not consume fermented nectar, so they put the nectar through a curation process to make it "shelf-stable." This final product is what we know as honey.

Is Honey Simply Bee Vomit?

You may have heard people refer to honey as bee vomit, and you even may have seen beekeepers who use this image in their branding. But there's so much more to the process!

Honey is highly concentrated nectar that has undergone important changes, and those changes start in the honey crop where the bee stores nectar during her flight back to the hive. The honey crop is critical to this process, because the forager's body adds two enzymes to the nectar. These two enzymes, invertase and glucose oxidase, start a chemical reaction and produce gluconic acid and hydrogen peroxide. The gluconic acid is what makes honey acidic, and hydrogen peroxide, as you know from your medicine cabinet, has antiseptic properties. Both the gluconic acid and the hydrogen peroxide prevent bacteria and mold growth, which contributes to honey's long shelf life. This process is why we need bees to produce honey; otherwise, we could just dehydrate the nectar ourselves and take honey bees out of the equation altogether!

In order to cure, or preserve, the nectar, the honey bees must dehydrate the nectar so that its water content is 18.5 percent or less. These receiving bees do this one of two ways: by holding a drop of nectar in their mouthparts until it is dehydrated or by fanning the combs containing nectar. Honey bees can control temperature and humidity in their hive, and this skill set is critical for dehydrating the nectar. Honey is what is called hygroscopic, which means that when it is exposed to air, it draws moisture from the air. Therefore, once the nectar is ripened into honey and the moisture level is 18.5 percent or less, the bees place a thin layer of beeswax over the top of the honey to prevent exposure to air. Think of this like preserving and canning fruits and vegetables. This process allows the bees to store food that can be consumed for many months or even years.

Who consumes honey and nectar in a colony? Developing worker and drone larvae, after their initial three-day diet of royal jelly, will be switched to a diet of bee bread and nectar. Honey can be fed to the developing larvae, but the worker bees will dilute the honey with a bit of nectar first. The biggest consumers of the honey and nectar stores, though, are adult worker bees. The carbohydrates found in nectar and honey provide the energy needed for all their daily activities. Larger colonies have more mouths to feed and more activity to support and therefore will require more carbohydrates.

The Effects of Poor Nutrition

Without the proper nutrition derived from these macronutrients of carbs, proteins, and fats, the colony as a system begins to fail. Poor nutrition negatively affects brood development, can shorten the length of the adult bees' lives, and can weaken the colony's immune system. A colony lacking nutrition is more susceptible to disease and may even begin to cannibalize its own brood to meet its protein needs.

It's important to note that a single floral source cannot provide all the nutrients a colony needs. In particular, a diverse supply of different pollen is important, as all pollen is not created equal. Pollen from different plant species will contain different levels of protein and varying levels of omega-3s and omega-6s. Even the sugar concentration in nectars of different floral sources can vary widely, from 5 percent to 75 percent or more. The same is true for humans, which is why nutritionists advise us to eat a variety of foods. Broccoli is very good for us, but a diet of *only* broccoli would lead to some serious deficiencies.

Hopefully, the natural environment will be able to meet most of your colony's nutritional needs. However, there will likely be times when this isn't possible. Pollen- and nectar-producing plants rely on the right weather conditions to grow, and this scenario provides a lot of opportunities for something to go wrong! For example, floods, freezes, or prolonged heat and droughts will affect when or whether plants grow and can produce the pollen and nectar honey bees rely on. Fortunately, a beekeeper can provide supplementary feeding replacements for both pollen and nectar if a colony is unable to meet its nutritional needs naturally. We will discuss the ins and outs of feeding bees in chapter 7.

pollen bee bread royal jelly nectar honey water

THE BEE'S SUPPER

CHAPTER 4

Getting Started Beekeeping

This is what you came here for—to become a beekeeper! If you've made it this far, you now have a good understanding of bee biology, nutrition, and how they construct and organize their homes. You are officially ready to be a good honey bee steward. Before we get into buying equipment, tools, and bees, though, let's learn more about the laws, costs, and time involved in this hobby to make sure it's right for you and your family. Beekeeping is incredibly rewarding and addicting, but it can get expensive quickly. I encourage you to be thoughtful about all the considerations up front before you dive in. I'm here to set you up for success.

And a warning about safety: though, generally speaking, honey bees are incredibly docile and there is no reason for people to fear bees, beekeeping is not without risk. We mitigate this risk through the use of certain tools and protective equipment. But for those that have a serious allergy to bees and are at risk of anaphylactic shock, I caution them to reconsider beekeeping as a hobby. If you are unsure whether you have a serious allergy to honey bees, please get tested by your doctor to confirm before you become a beekeeper. Also, I strongly recommend that you carry a first-aid kit with an ice pack, some oral antihistamine, and—if your doctor recommends it—an epinephrine injector.

Beekeeping Laws

First, it's important to explore local beekeeping laws and regulations. Beekeeping has become a popular hobby and is legal in many urban areas. But before you start buying equipment and bees, research your city, state, and county regulations on apiaries. (An apiary simply refers to a collection of honey bee hives.) If you are unsure where to look, calling local and state departments of agriculture is a good place to start. Also, local beekeeping clubs can be very helpful. Municipalities may limit the number and type of hives allowed, require fencing, or require that hives be a specific distance from a beekeeper's property line. And don't forget about homeowner's associations; many of these associations ban backyard beekeeping outright.

Time Investment

Many are relieved to hear that honey bee hives do not require daily maintenance, while others are disappointed when I tell them how infrequently they should actually be inspecting their hives. You will learn more about hive inspections later, but the frequency of hive inspections will be shaped by a beekeeper's philosophy and whether or not there is an issue they are monitoring in an apiary. Generally speaking, I recommend that beekeepers aim to conduct hive inspections once every two to three weeks over the course of the year. Inspections should occur more frequently during the warmer months and less frequently, if even at all, during the cold winter season. Beekeepers with hives located in areas with very long winters may not inspect colonies for several months at a time. The time spent inspecting hives isn't a deterrent to starting beekeeping for most people, but beekeepers will spend far more time learning, reading, and trying to understand their hives than actually inspecting colonies. Beekeeping is a commitment to lifelong learning! You will learn more about the seasonality of beekeeping in chapter 12.

Choosing a Hive

One of the first critical decisions you must make before becoming a beekeeper is what type of hive you plan to keep. In chapter 2 you learned about how a honey bee colony builds and organizes its hive in the wild. Keep in mind that the organization of a colony will be the same in any human-made vessel as in the wild. The difference is that the human-made hives have removable frames or bars so that the colony can be inspected without damaging the hive.

Beekeepers have several types of hives to choose from, but the two most common for hobbyist beekeepers in the United States are the Langstroth hive and the *top bar hive*. Beekeepers tend to be very opinionated about the "right" way to keep bees, and you may receive pressure to do one over the other. Both hive types work well, and aside from a few differences in tools and techniques, the gist of the beekeeping is the same in each. I started my first apiary with top bar hives but moved to Langstroth hives when I started my honey business. Let's examine the primary differences between the two.

ANATOMY OF A HIVE

→ outer cover

→ inner cover

frames

LANGSTROTH HIVE

hive bodies
(aka brood boxes
or supers)

→ bottom board

TOP BAR HIVE

top bar

follower
board

LANGSTROTH HIVES

A Langstroth hive is the more common of the two hive types. A Langstroth (or "Lang") is a modular hive that consists of a series of stacked boxes known as hive bodies. These hive bodies are four-sided boxes without a top or a bottom. As the hive grows and the bees require more space, the boxes can be stacked on top of one another, expanding the space for the bees to live and allowing them to move freely between the hive bodies. A Langstroth hive comprises the following components.

1. **Outer, or telescoping, cover:** This outer cover protects a colony from the elements and will likely have a metal top to help deflect heat and assist with insulation.

2. **Inner cover:** An inner cover goes between the outer cover and the upper-most hive body in the hive. It will have a hole in the middle and helps with ventilation and moisture control in your hive.

3. **Hive bodies:** Begin with one hive body initially, and add more as the colony grows and needs more space. Hive bodies will contain four-sided frames, which is where the bees will build their beeswax comb. The hive bodies also come in different heights: you can purchase what are known as deep (9 5/8"), medium (6 5/8"), or shallow (5 11/16") hive bodies. Hive bodies can also be called brood boxes or honey supers. These terms refer to what is contained (or what a beekeeper expects to be contained) inside of these boxes. Hive bodies that house the brood nest are called brood boxes, and hive bodies that contain the honey stores above the brood nest are called honey supers.

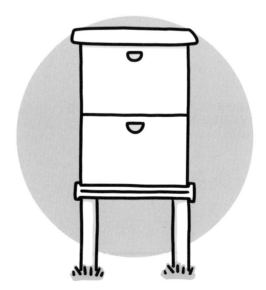

4. **Frames:** Each Langstroth hive body will contain frames. Frames are rectangular and can be made from wood or plastic. The bees will build their beeswax comb and attach it to the top and sides of the frame, continuing to build out hexagonal cells until they have filled up the frame. It is in this beeswax comb where the queen will lay eggs and the foraging bees will store food. When bees fill up the space with beeswax comb, this is called drawn comb. Frames are available either with or without foundation. Foundation is a sort of "starter" sheet made of beeswax or plastic that hangs inside the frame. The honey bees will build their beeswax comb on both sides of this foundation. The foundation gives the comb a bit of structural reinforcement so that it's not as susceptible to breaking or drooping when inspecting a colony, and foundation can be helpful when harvesting later. If choosing foundation, I recommend opting for either beeswax foundation or plastic foundation coated in beeswax. Honey bees tend to avoid building directly on unwaxed plastic foundation. Frames also can be used without foundation, aptly called foundationless frames. Foundationless frames must include some guide at the top to help ensure that bees build straight comb. In our yards, I prefer to use a mix of foundationless and some beeswax foundation frames. You might want to experiment with all types and see which you (and your bees!) prefer.

5. **Bottom board:** A bottom board is the floor of the Langstroth hive. The front of the bottom board contains the hive entrance and is where the bees will enter and exit the hive. As discussed in chapter 2, solid and screened bottom boards are available, but I prefer to use solid bottom boards in my apiaries.

The greatest benefit of the Langstroth hive is that it is a standardized system. That means parts bought from different suppliers will fit together. Langstroth hives also provide the opportunity to manage space by giving and taking away hive bodies as a colony grows or contracts over time. The greatest detriment to using a Langstroth hive is that it requires bending and lifting of hive bodies in order to inspect the hive. When these hive bodies are filled with honey, they can be quite heavy.

Langstroth hives come in two widths: they are sold in eight-frame or ten-frame configurations. This designation refers to the number of frames that each hive body holds. This is a personal choice that doesn't much affect the bees. If you are concerned about the weight of the boxes, choose an eight-frame configuration. Choosing a ten-frame configuration may save you a little bit of money because you get more space per hive body, and it also may mean you can get away with using one fewer hive body as your colony grows.

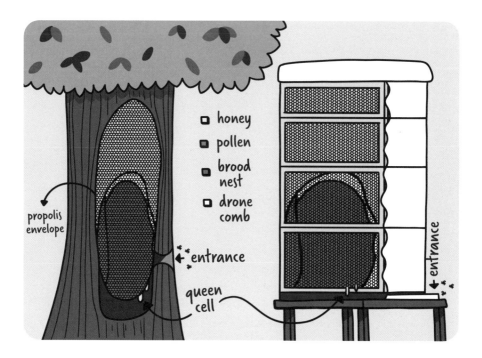

It can be difficult for someone new to beekeeping to envision how a colony would use the space in a Langstroth hive. I remember seeing diagrams of a Langstroth hive before I got my first colony of bees and not understanding where the brood, honey, and pollen would go inside the hive. Return to chapter 2 to refresh your memory of how a colony organizes itself in the wild. This is exactly how it will configure itself in the hive. The only difference is that the human-made hive will have removable frames. Above, I've included the same illustration of a wild colony from chapter 2 and demonstrated how that same colony will look inside of a Langstroth hive.

TOP BAR HIVES

A top bar hive is composed of a long, singular box. Inside the hive, bees attach their beeswax comb to top bars, bars that sit across the top of the long, singular box. The bees will build comb on these bars along the horizontal axis, from the front to the back of the hive. A top bar hive also may include what is called a follower board, which allows a beekeeper to limit the space available to the bees. The entrance to the hive is usually in the form of one or more holes drilled into the long box.

The greatest benefit of the top bar hive is that it usually sits at waist height and doesn't require any bending or lifting of hive boxes. However, great care must be taken to ensure that the bees build straight beeswax comb along the top bars. Unlike the rectangular frames of a Langstroth hive, these top bars have no sides, bottom edges, or foundation to guide the shape of the comb. If the bees are allowed to build across the bars (called cross-combing), a top bar hive can become unruly very quickly, and the top bars cannot be removed to inspect the colony without damaging the comb. The top bars of beeswax comb must also be handled very carefully. Because the comb is attached only on one side and not on three or four sides, as is the case with a Langstroth frame, these top bars must be kept perpendicular to the ground at all times. Even a slight tilt on the axis of the top bar can cause the beeswax comb and its contents to fall from the bar onto the ground. Also, unlike in a Langstroth hive, a top bar hive has a finite amount of space, which means your bees must be managed very carefully to prevent swarming. You will learn more about swarming in chapter 8.

Also, unlike Langstroth hives, top bar hives are not standardized. This means that top bars from hives purchased from two different suppliers likely will not be interchangeable between the hives. You will learn in chapter 5 that the ability to share resources between hives is critical, so if you choose to buy top bar hives, make sure to purchase from the same supplier so that all the hive components are compatible. Many people who want top bar hives choose to build their own, and free plans can be found online.

The Evolution of the Modern Beehive

The relationship between honey bees and humans dates back many thousands of years. Ancient Egyptians believed that honey bees were created when tears from the sun god, Re (or Ra), fell to the earth. Before ancient peoples discovered that honey bees could be kept in vessels and managed as livestock, they participated in honey hunting. Honey hunting is seeking out wild honey bee colonies and robbing the colonies of the honey they have produced. The earliest depiction of honey hunting appears in a cave painting in the Cuevas de la Araña of Eastern Spain and is estimated to be eight thousand years old.* The hunter is reaching into a wild bee colony, taking honey, and placing it into a basket. Honey hunting is still practiced to this day, including in Nepal, where hunters scale high cliffs in search of honey.

Once humans realized that it was much easier to catch or lure bees into vessels for honey production, many different types of traditional hives came into use. The first evidence of humans tending to honey bees in artificially made cavities, such as clay pots, is about five thousand years old.† The artificial hive types varied by culture, including skeps, which are hives made from mud, clay, straw, or wicker woven into what looks like an upturned basket. Skeps made from straw and wicker were widely used throughout Europe into the late 1800s. Skeps are commonly depicted in children's books and in other bee and beekeeping imagery. All of these traditional hive types are now illegal in many areas of Europe and the United States, because without removable frames the hive cannot be safely inspected for pests and disease, and harvesting the honey entails destroying the colony.

The earliest hives with removable comb can be found in top bar hives. Evidence suggests that early beekeepers in Greece would lay sticks across the top of a pot or basket. Top bar hives are the primary hive type used in many developing countries, because almost any vessel can be turned into a top bar hive, including barrels and empty tree trunks. They are simple to build and, unlike a skep, don't require killing the hive for harvests.

In 1851, Reverend Lorenzo Langstroth built on other beekeepers' early designs of hives with removable frames and invented what is now known as the Langstroth hive. His design recognized the rules of bee space (see chapter 2 for a reminder), and he created a hive with frames that could be removed, replaced, or moved to other hives without damaging the comb. It is Langstroth's design that is the most commonly used hive type in the Western world.

* The paragraph up to this point is borrowed from Gene Kritsky's *The Tears of Re: Beekeeping in Ancient Egypt* (Oxford, England: Oxford University Press, 2015).
† Kritsky, *The Tears of Re*.

OTHER HIVE TYPES

Though top bar hives and Langstroth hives are the most common hive types today, beekeepers do have other options. You may find one of these other hive types is better suited to your preferences and needs, but they are less common, so there may not be as much documentation of best practices for working with them. Following are just two other hive types beekeepers choose.

Warre Hive

A *Warre hive* (pronounced *WAR-ray*) is named after its founder, a nineteenth-century French monk named Émile Warré. It acts as a sort of compromise between a Langstroth hive and a top bar hive. Like the Langstroth hive, the Warre hive has a modular design with a series of vertically stacking boxes. But instead of rectangular frames, it has top bars that sit across the top of each box. A Warre hive is designed such that boxes are added at the bottom of the hive instead of stacked on top of the hive like a Langstroth. In a Warre hive, the bars do not have four sides and cannot be used with foundation. Therefore, similarly to a top bar hive, great care must be taken to ensure that the bees build straight and to preserve the integrity of the comb during inspections.

Long Langstroth Hive

Often referred to as a long Lang, this hive is situated along a horizontal axis like a top bar hive but uses standard deep Langstroth frames. As with a top bar, the *long Langstroth* relieves a beekeeper from bending and lifting and has the added benefit of standardized use of Langstroth frames, with or without foundation.

A FEW BEST PRACTICES

Whatever you choose, I recommend that you select only one hive type for your apiary your first year. This will not only allow you to focus your learning on one style but also give you the ability to share resources between the hives to help boost weak hives or those low on food stores. You will learn later that pollen, bee bread, nectar, honey—even brood—can be shared among honey bee colonies.

For this same reason, I always encourage new beekeepers to consider waiting to start an apiary until they can afford to have at least two hives. Starting with just one hive means a beekeeper does not have the ability to share resources if a colony becomes queenless, grows weak for a number of reasons, or is light on honey. Having at least two hives also allows a beekeeper to compare the two in terms of behavior, strength, and more.

Once you have decided on your hive type, it's time to order honey bees! You can hang swarm traps to catch a swarm of bees (more on this in chapter 8) or buy a fully established colony, but many new beekeepers who start new hives choose to buy bees from bee breeders.

Two options are available: a *package of bees* or a *nucleus colony*, known as a "nuc" for short. Let's examine each of these options:

A PACKAGE OF BEES

A package of bees is sort of like a swarm in a box. The beekeeper receives a ventilated box with thousands of worker bees and a queen bee. Packages of bees are measured and sold by weight. Weights can vary, but three pounds, or roughly ten thousand bees, is fairly standard.

To create a package of bees to sell, the bee breeder shakes bees from different hives into this screened box and then adds a mated queen. A mated queen means she is ready to lay eggs once she's in her new hive. Because these bees come from different hives with existing queens, they need some time to accept their new queen. Therefore, a package of bees will come with a queen bee in a separate, small screened box called a *queen cage*. This queen cage allows the bees in the package to detect her pheromones while providing her protection from their stings. (Recall from earlier that bees will kill an unfamiliar queen bee.) It will take only three to five days for the worker bees to adjust to her pheromones and accept her as their new queen. Also within the ventilated package box will be a food source, usually a can of sugar water, that the bees can access for food.

A package of bees needs to be installed into a hive as soon as possible—no more than a day or two after they are picked up. A package of bees is installed by shaking the bees into a hive and hanging the queen cage between two frames or top bars using either the clip attached to the queen cage or a zip or twist tie. You can find videos of package installs on our YouTube page!

Queen cages may vary, but most will have a hole in one end that is plugged with a hard piece of sugar. As the attendants eat the sugar plug over several days, they will eventually release the queen into the hive. The other bees hopefully have now acclimated to her pheromones. If your queen cage has a plug preventing the bees from consuming the sugar, be sure to remove this plug before installing. But do not remove any plugs that will immediately release the queen! Then, during a future inspection, you can simply remove and discard the now-empty queen cage.

A NUCLEUS COLONY

Think of a nucleus colony, or nuc, as a mini version of a regular colony. It will include anywhere from four to six frames or top bars of beeswax comb with developing brood, honey, and pollen already in the comb, along with nurse bees and a laying queen. To make a nuc, a bee breeder removes a few frames or top bars of brood, honey, and pollen from existing colonies and places them into a small box. Then the bee breeder provides this new colony with a queen bee or allows the bees to rear their own. (This process is also called making a split. You will learn more on when and how to make splits in chapter 9.) Presuming your nuc box has an entrance that you can open, you don't need to install the nuc into your hive right away. Instead, place the nuc on top of your hive and open the entrance to allow the bees to start to orient themselves to their new home and begin foraging for food.

To install the nuc, simply remove the frames or bars from the nuc box and place them directly into the hive. There will likely be some bees left crawling around the inside of the nuc box. Once you have transferred the frames or top bars, shake the remaining bees into the hive and leave the nuc box next to it so the stragglers can find their way to their new home by following the phero-mones of the queen and their sister worker bees.

There is one important difference between packages of bees and nucs: unlike a package of bees, the queen is *not* in a separate small box in a nuc colony. She will be mingling among the thousands of worker bees, and it is critical that she make it from the nuc box to the hive. As you transfer over the frames or top bars, look at each and see if you can spot her. Since this may be your first time queen spotting, you may not find her. In that case, make sure you get as many of the bees out of the nuc box and into the hive as you can! Be sure to install the nuc into your hive within a few days, as the growing colony can quickly run out of room.

CHOOSING BETWEEN A PACKAGE AND A NUC

Here are a few things to keep in mind when deciding whether to buy a package of bees or a nucleus colony.

First, know that packages are not available in all areas, and their availability is more seasonal than a nuc. The bees in a package have a lot more work to establish the colony because they are not coming with their own food and drawn comb, and therefore are only available in the springtime. For this reason, packages of bees usually are not available in areas with long winters and short bee seasons. That said, packages are the most flexible option, as you can install one into any type of hive. Packages are also generally a more affordable option.

Second, if you purchase a nuc you must ensure the nuc type corresponds to the hive type that will house the nuc; otherwise, your frames or top bars won't fit into your new hive. For example, if you buy a top bar nuc, you must make sure that your top bar hive has the same dimensions as the nuc you purchase. And when it comes to Langstroth hives, almost all nucs are created for deep-size hive bodies, so you will need to make sure you have at least one deep hive body to accommodate those larger frames.

Third, if you are purchasing bees, look local. Unless you live in Europe, Asia, or Africa, honey bees are not native to your area. Therefore, finding a bee breeder in your region will help ensure the bees are well adapted to your climate. I also recommend researching your options carefully to learn a bit more about what practices any breeder employs, such as what treatments or antibiotics they use, and to ensure they select for hygienic behavior and gentle temperament. Working with a local bee breeder is also an excellent way to grow your beekeeping network. They'll have information and tips that will be valuable to any beekeeping practice.

There are dozens of different subspecies of honey bees. For example, European subspecies include Italian, Russian, and Carniolan honey bees. Although many beekeepers praise the traits associated with different honey bee subspecies, the likelihood of receiving a true purebred queen is likely very rare and unnecessary. First, thanks to globalization, honey bees have interbred in the United States for many years. Second, even if you do find a purebred queen bee, her offspring only receive half of her genetics. Queen bees mate with many drones, which can be controlled by a beekeeper only if they artificially inseminate their queens. Therefore, almost all the bees in areas where honey bees aren't native, such as the United States, are mixtures of multiple stocks.

Rather than looking for these particular subspecies, focus on the breeder's practices, values, and ability to select for ideal traits. In my own apiaries, I buy bees only from breeders whose queen bees exhibit a trait known as *varroa sensitive hygiene* (VSH). VSH is a trait that allows bees to detect and remove pupae that are infested with the parasitic mite *Varroa destructor*. You will learn more about this mite in chapter 10.

Also, keep in mind that bees are not available for purchase year-round. I recommend looking to preorder your bees in the fall or winter season for delivery in the spring. Finally, be sure your apiary site and hives are set up and ready on the date the bees are available for pickup. Packages need to be installed as soon as possible—preferably the same day—and nucs should be installed within a few days.

Questions for Bee Breeders

Choosing a breeder can be daunting and overwhelming. You may have many options in your area or none at all. For those who have several options to choose from, I have provided a list of questions to ask any breeder before making a purchase.

GENERAL QUESTIONS

1. Who supplies your queen bees for your nucs and packages, and where are they bred?
2. Have you been recently inspected by the state apiary inspector? What was the result of that inspection?
3. Is the stock that produced the queens treated for mites? If so, how often, and what treatments are used?
4. What traits do you breed for in your breeding program?

Some examples of desirable traits include:
   a. Disease resistance
   b. Hygienic behavior
   c. Gentle temperament
   d. Honey production

BUYING PACKAGES

It's important to understand how many days ago the beekeeper prepared, or "shook," your packages before they were made available for pickup. Packages of bees are extremely perishable, and it is critical that they are installed as soon as possible. On your pickup date, inspect the package of bees. Though some dead bees on the bottom of the package is expected, it should be minimal. To learn more and watch videos demonstrating installing packages of bees, check out the Two Hives Honey YouTube channel. Some questions specific to packages include:

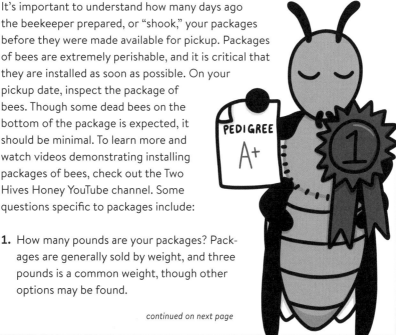

1. How many pounds are your packages? Packages are generally sold by weight, and three pounds is a common weight, though other options may be found.

continued on next page

2. How many days before pickup did you prepare or shake the packages?
3. On your pickup date, inquire about the last time the bees were sprayed with sugar water. If it has been more than twelve hours, be prepared to give them a quick spray to keep them cool and give them a snack! Use one part plain, white table sugar to one part water. A light spray is all that is necessary.

BUYING NUCLEUS HIVES

With nucs, it's important to understand what you are buying. This includes the number and size of frames and what they contain. Also inquire whether an inspection of the nuc is allowed before taking it home. You may or may not decide to inspect your nuc, but if you do, you want the nuc to match the description as it pertains to numbers of frames of brood and food. A nuc should not have any drone brood (though it may contain a small number of adult drones, and that is not a concern). Every frame should be covered in bees. To learn more and watch videos about installing nucs, check out the Two Hives Honey YouTube channel. When selecting a nuc, ask the seller:

1. How many frames are your nucs?
2. How many of the frames are brood and how many are food (nectar/honey/pollen)?
3. Are all the frames drawn out and full with either brood or food, or is one empty frame provided to "grow on"?
4. If the nucs are Langstroth hives, does the nuc contain deep or medium frames?
5. If the nucs are top bar hives, what are the measurements of the top bars and drawn comb? (Remember that top bar hives are not standardized, so check that the top bars will fit into your hive.)
6. Have you confirmed that the queen in the nuc is laying eggs?
7. May I inspect the nuc before accepting them?
8. Is the nuc box included in the price, or am I expected to return it?

Tools and Protective Gear

There is no shortage of beekeeping gadgets and gizmos. Flip through any beekeeping catalog to find thousands of items marketed to beekeepers. I'm a minimalist and will share what tools are a must for beekeepers.

Hive setup: Each colony in your apiary will need a number of components to complete the hive. For a Langstroth hive, this will include an outer cover, inner cover, bottom board, at least two to three hive bodies, and enough frames to fill each of the hive bodies. This means eight frames in each eight-frame hive body and ten frames in each ten-frame hive body.

Each top bar hive should include the actual hive and hive top and enough top bars to fill out the hive, and it may include a follower board. You will learn in chapter 5 that colonies should not be given more space than they can adequately guard, and a follower board allows you to accomplish this task in a top bar hive.

Hive stand: Each hive should be on some sort of stand so it is not resting on the ground. Many top bars come equipped with a stand to house the top bar hive at waist height. For Langstroth hives, you can buy or make a hive stand or use something as simple as two cinder blocks. Do not place a Langstroth hive at waist height, like a top bar hive. This is because as the Langstroth hive grows vertically, you will have to lift heavy hive bodies at eye level and above. This is not advisable. Instead, use a stand that situates your Langstroth hive anywhere from eight to sixteen inches off the ground.

Entrance reducer: An entrance reducer is used to decrease the entrance size of a Langstroth hive.

Feeder: You will learn more about feeding bees and different feeder types in chapter 7. Plan to purchase one feeder for each of your hives.

Protective gear: Protective gear comes in a number of material types and configurations, and I strongly recommend a fuller coverage piece. This means either a jacket and a veil or a full suit and veil, made from ventilated material for comfort and improved sting resistance. As your comfort level grows, you may later decide to wear less protection, such as a veil only. However, every beekeeper should have a protective gear option that provides full coverage, head to toe. I most often wear just a veil, but I always have a very sturdy pair of oversize coveralls that I can toss on with my veil for more protection. Even if you need your full protection only one day a year, I promise it's worth the money!

Veils come in a number of different styles, which is a good reason to shop a local beekeeping supply store if possible. Trying different styles of veils will allow you to choose what you most prefer. Finally, make sure your protective gear fits you properly, and err on the side of larger if necessary. You want to make sure you can raise your arms over your head and bend over comfortably without the suit riding up or exposing parts of your body to the bees.

Next, you will need a pair of gloves. I prefer traditional leather beekeeping gloves that fit snugly, as the tight fit helps with dexterity. Any leather style will help protect from stings, but gloves that aren't made for beekeeping tend to have a cuff that ends at the wrist. A longer cuff, as is found in traditional beekeeping gloves, will provide more coverage in this area. Many beekeepers use nitrile gloves instead of leather gloves. Nitrile gloves are sting-resistant and allow for maximum dexterity. However, they don't allow the skin to breathe. They aren't for me, as I find they make my days in the apiary miserable, as I get overheated

WHAT'S IN A BEEKEEPER'S TOOLBOX?

AN ENTRANCE REDUCER

HIVE STAND

OR

CINDER BLOCKS

A FEEDER

(see types in ch. 7)

A SMOKER

A BEE BRUSH

A HIVE TOOL

A FRAME STAND

PROTECTIVE GEAR

ore quickly while wearing them. However, plenty of experienced beekeepers love them. I encourage you to try both and see what works best for you!

Hive tool: Think of a hive tool as a sort of pry bar to break open the propolis seals in a hive. The propolis seals on an established hive are almost impossible to break without a hive tool, which is why I recommend buying several to have on hand in case you misplace one.

Bee brush: A bee brush is just that: a brush used for bees! Bee brushes are most useful during harvesting honey and for getting bees off your body before you take off your bee suit. Be warned that bees aren't particularly fond of getting brushed. Because of that, I rarely use them to remove bees from frames of bees and instead will use a swift downward motion to shake the bees off the frames.

Bee smoker: A smoker is critical for the safety of the bees and the beekeeper during a hive inspection, and you will learn more about this important tool in the next chapter. Bees communicate via pheromones, and the smoke from the smoker masks the alarm pheromone given off by the guard bees at the entrance of a hive when they feel threatened. This same pheromone is released when a bee dies after stinging. A quality bee smoker can last many years. The bellows will likely need replacing from time to time and can be easily replaced using just a screwdriver.

Comb or frame stand: A comb stand isn't required, but I do find them to be very helpful during inspections. A comb stand attaches to the side of a hive and allows space to hang frames or top bars as a beekeeper works a hive. Removing a few frames or top bars to allow some space makes inspections a bit easier, and a comb stand provides a safe place to hang them.

Estimating the Financial Investment

The tricky part about the financial investment involved with becoming a bee-keeper is that the two most expensive elements, hives and bees, are necessary to become a beekeeper. If you wanted to try your hand at becoming an artist, you could buy a few notepads and inexpensive paints or pencils and see how it felt before investing in more expensive tools and supplies. Not so with beekeeping.

That said, many of my most successful students started as just that: students eager to learn about bees. I recommend folks find ways to get involved with a local beekeeping community before buying bees. Local beekeepers may provide classes, and there might be one or more beekeeping clubs with folks eager to mentor aspiring beekeepers. Finding some way to get into honey bee colonies with another beekeeper is a great way to see if you are still excited about making a more serious financial investment!

Lessons from a Beekeeper

When it comes to beekeeping tools and gear, literally thousands of items can be found in beekeeping equipment catalogs. Over the years, I've learned where it's OK to save and skimp a bit (($)) and where it's better to spend more to ensure quality (($$)).

($$) HIVE COMPONENTS: You can find cheap all-in-one kits online, but they often come with unnecessary items while omitting some necessary components. I strongly recommend buying hive components à la carte to ensure you have all the items you need and the type you most prefer. For example, the feeder that comes with the kits may not be the feeder type most recommended or best suited for you. Most hives found at beekeeping stores will be made from pine, though some suppliers sell sturdier options made from cedar at a higher price. If choosing wood for the hive components, be sure to invest in a good outdoor paint or sealant and *paint or stain the outside of the hive only* to ensure that you get the maximum life out of the woodenware. Unpainted or unfinished pine hives won't last long in the elements. Avoid painting or staining any of the interior walls of the hive. If you or someone you know has carpentry skills, you can find free plans for different hive types online. In this case, choosing a higher quality wood than pine, if price allows, can help ensure the hives last as long as possible. In recent years, hives made out of polystyrene, a type of plastic, have become more popular and available. The benefits of these polystyrene hive components are that they are much lighter than their wooden counterparts and they provide much better insulation for the colony. The trade-off is that they are usually much more expensive than pine components. Some folks are also turned off at the thought of using nonorganic materials, such as polystyrene, to house their bees and the environmental impact of using plastic. A compromise we use in some of our yards is to use polystyrene top covers on top of our wooden bee hives. One study found that half of the heat in a colony is lost through the top,* so this gives our bees significantly more insulation where the colony is losing heat most—through the lid.

($$) PROTECTIVE GEAR: Protective gear comes in all sorts of fabrics and materials, from canvas to cotton to mesh. I strongly recommend purchasing the highest quality protective gear you can afford for yourself. This is, after all, your primary protection against bee stings. Bees can sting through a single layer of cotton or canvas if it is pressed up against the skin, but the more expensive, ventilated gear provides three layers of protection so the stinger can't reach skin as easily. It's also much cooler and more comfortable in the heat. If you aren't well protected, you may be more focused on the potential for stings than on your hive inspection! Save the cheaper gear for any guests who want to visit your apiary. This gear won't get near as much use as your primary protective gear.

$ HARVESTING EQUIPMENT: Extraction tools and gear can get pricey quickly, and most beekeepers will not be harvesting honey their first year. If you are one of the lucky few who get to harvest in the first year, there are a number of tools you can find in your kitchen to help process your first harvest. Push the cost of the harvesting equipment off until later, when you can better assess your needs.

$ HIVE STAND: Hives should be at least eight inches off the ground. More expensive hive stand options are available, but cinder blocks work well and can be procured inexpensively or for free.

$$ SMOKER: Similar to protective gear, smokers are available at all quality levels. Also similar to protective gear, a smoker is critical to your ability to protect yourself while working your hives. Don't skimp here. Choose a smoker with a quality bellows, preferably one with a cage that extends down below the bottom of the smoker. This cage helps keep the hot metal away from anything that may catch fire or burn easily (including skin!).
A quality smoker can last your entire beekeeping career, and only the bellows will need replacing from time to time. I also recommend a box or bucket made from metal to store the smoker in after use to help contain any embers flying from the smoker.

$ OTHER MISCELLANEOUS TOOLS: A hive tool and a smoker are the only two tools I always have in hand when I'm in an apiary. Most other "must haves" can be improvised. Unless you are really into gear or aren't on a budget, save money on the purchase of other tools the first year.

* Daniel Cook et al., "Thermal Impacts of Apicultural Practice and Products on the Honey Bee Colony," *Journal of Economic Entomology* 114, no. 2 (April 2021): 538–546.

Costs for hives and bees can vary wildly from one region to another and also will depend on whether you build or buy the hives, the type of hives purchased, and the method used to procure bees. If you are price sensitive and budgeting is important to you, keep reading for tips on how to minimize costs and to learn where it's OK to skimp and where it's wise to go for quality items, even at a higher price.

- **Hives and accessories:** For each colony you want to start, you will need to buy or build the necessary hive components. If keeping to a budget is important, research this cost first, as it will make up half or more of the costs associated with each hive. Each hive will need a feeder and a stand so the hive is not sitting directly on the ground.

- **Bees:** Later you will learn how to catch a swarm of bees, and if you are successful in this endeavor you don't need to budget for this cost. However, for most new beekeepers buying a package or nucleus colony of bees from a breeder is the easiest way to start. As with the hive components, bees will make up a substantial part of the cost, so include this in your initial budget. Each hive you want to start needs one package of bees or a nucleus colony. I recommend starting with two to four hives.

- **Protective gear:** Every beekeeper who plans to work in the apiary will need a set of protective gear. It's also not a bad idea to have an extra set of gear on hand. Once you have bees, you will likely have lots of requests by visitors and friends to join you in the apiary! If possible, plan to purchase these items in person so you can try them on. If not possible, plan to size up one size in the gear. Part of what makes the suits effective is that the loose-fitting gear helps prevent stingers from reaching your skin. You won't experience many negative effects if the gear is too big, but if a jacket or suit is too small, it is going to be uncomfortable and can allow bees into your gear when you bend or lean over.

- **Tools:** There are a number of tools that can be purchased once and rarely need replacing. For budgeting, I support and advocate for a more minimalist approach to gear and tools. A beekeeper really only needs two tools to work bee hives: a smoker and a hive tool. Other tools can always be added later. One other tool to consider is a comb stand, which provides a safe place to place frames during hive inspections.

- **Apiary setup:** Depending on the site you select, you may have a number of costs associated with preparing the site. You will learn more about best practices for setting up the apiary site shortly, but this may include fencing if you have livestock or live in an area with bears, as well as costs for leveling the area and ensuring that no foliage grows within at least a few feet of each hive.

A final note about purchasing tools and gear: whenever possible, I encourage you to support your local beekeepers and beekeeping businesses, as they will be a wealth of knowledge and resources for you. If your area does not have any local beekeeping shops, you can find all these tools and more online.

Setting Up the Apiary

You've selected your hive type and ordered your bees, and now it's time to start to plan your new apiary! When selecting an area for the bees, keep a few things in mind:

1. Be sure to follow any local and state regulations in selecting a site, as the hives may be required to be a certain number of feet away from your property line.

2. A site with at least three to four hours of full sun is most advantageous for the bees. Sun helps control for small hive beetles, a hive pest, and helps with ventilation of the hive. Morning sun is most ideal, as the sooner the sun hits the front of a hive, the quicker the bees get the message that it's time to begin foraging for food each day. For this same reason, facing a hive's entrance toward the south/southeast is best, but is not a requirement if this doesn't work for the space.

3. Be sure to select a site where no one will be working or hanging out often in the hive's bee line, the eight to ten feet of space directly in front of the hive's entrance. Anyone ever in this space will want to be protected with gear and a bee smoker, so site the hives accordingly around your daily activities.

4. Level the site. Though you don't need to get out an actual level for this exercise, as close to level as possible is ideal, particularly if you have a top bar or aren't using foundation in your Langstroth frames. Bees will always build their comb perpendicular to the ground, so ensuring your frames or top bars are in this configuration is critical to making sure the bees build straight comb. Having the hive tilted *slightly* toward the entrance will allow any water, which may collect after a rain, to run out of the front of the hive easily. The back of the hive should never be lower than the entrance, as this will allow rainwater to pool in the back.

5. The site should also be on solid, level ground where pooling of water does not occur. Shifting earth can cause hives to settle unbalanced and topple. Hives also should be elevated at least eight inches off the ground, and if you are using Langstroth hives, ideally no more than sixteen inches. Any higher may require lifting of heavy boxes over your head as the hive grows! Elevating the hives helps with ventilation, deters pests like raccoons and

possums, and protects the hives from flooding waters. Plus, keeping the hive off the ground extends the life of the hive components.

6. Once you've selected a site, make sure that lawn maintenance won't need to occur within at least four or five feet of the hive. Weed cloth, stones, mulch, or paving stones can be used to prevent growth, as can a concrete pad. This will prevent foliage from interfering in your ability to inspect the hives. Also, avoid any sort of lawn maintenance or weed eating right next to the hives. If machinery such as a lawn mower or weed eater must be used near the hives, be sure to smoke the hives first, keep a smoker on hand as needed, and wear protective gear.

7. If you have livestock or live in an area with bears, you will want to consider fencing. (And in the case of bears, electric fencing may be required!) Livestock and bees can coexist happily, but livestock also have a tendency to knock hives over.

8. Finally, ensure that the bees have a water source. Bees need water for hydration, to cool the hive, and to rehydrate honey to feed to developing larvae. A few things to keep in mind: bees drown very easily, so some sort of landing pad is important. Rocks or floating wine corks work well. Also, I find bees tend to prefer water a short distance (about fifteen to thirty feet) away from their hive. If you or a neighbor have a swimming pool or livestock tanks, giving the bees a water source will be critical to ensure they don't visit places where they aren't welcome. Place the water source between the hives and the area you want them to avoid. Adding a few drops of lemongrass essential oil to the water can make it more appealing and help ensure they choose the provided water source instead of a swimming pool.

Lessons from a Beekeeper

If beekeeping is your first foray into animal husbandry and agriculture, buckle your seatbelt. . . . It's a wild ride! Beekeeping can be both incredibly rewarding and heartbreaking. Because of a colony's ability to rear new queens that can then continuously produce new worker bees, a honey bee colony can theoretically be both immortal. But you will experience losses. The first is always the hardest, but it certainly doesn't get easier, at least for me, to lose colonies. The most troublesome losses are those where I can point to how poor decision-making, planning, or even neglect on our part caused a loss. Following is a case in point.

A client, let's call him Chad, had picked out a site for a new apiary near a creek bed at a property. I expressed my concern about water levels and potential flooding from the creek. Chad assured me it had not flooded in over sixty years in this area. I also expressed my concern that any heavy rains would make it challenging for my beekeepers and me to get our trucks up and down the steep rise leading to the apiary, and he replied that no one had ever gotten stuck on this incline. Pushing aside our concerns, we set up the apiary based on Chad's recommendations. Two months later, Sarah, one of my Two Hives Honey beekeepers, called me from the yard. We had just received one of the heaviest rains in recent memory, and she had bad news.

First, she was stuck at the bottom of the rise. And second, the apiary had disappeared. It was completely gone! She hiked down the creek bed and found remnants of the hives all along the way . . . a hive body here, a cinder block there. The heavy rains had caused the creek bed to swell, and it washed away all four hives. I was physically sick to my stomach, knowing we had ignored our instincts and lost four colonies in the process. With a heavy heart, I picked up the phone to call the client's assistant and inform him of the devastating news.

After I described what had happened, the assistant broke into hysterical laughter. I was completely caught off guard by the reaction. Through fits of giggles, he said, "It never fails; everything Chad touches turns into a mess." This unexpected reaction caused me to laugh hysterically too. A good thirty-second fit of communal (albeit inappropriate) laughter passed, and through tears I shared potential next steps. Unbeknownst to him, this gentleman did teach me an important lesson that day. If you're not ready for the heartbreak, you're not ready for beekeeping. But the losses and heartbreak are part of the package deal of the delight of beekeeping. Learn your lessons, trust your instincts, and do better next time. And take the time to laugh at yourself along the way. Also, never ever put hives on a creek bed, no matter what Chad says.

CHAPTER 5

Performing a Hive Inspection

By this point, you've put a lot of work into getting your apiary set up. You've selected a hive type, procured or built a hive, found a breeder and ordered bees, prepped the apiary site, and more. But once those bees are actually in the hives, *then* what? How much care do they need, and how frequently do they need it?

Let's make one thing clear up front: honey bees don't need humans to survive. In fact, an inexperienced beekeeper's interventions can do more harm than good, usually from a lack of knowledge about what colonies need and don't need. Recall from chapter 1 that wild honey bees have existed and thrived without human intervention for millions of years. However, if you are choosing to keep bees in your backyard, you presumably have invested significant time and money in this endeavor, which means you want to protect that investment. Plus, you may have an interest in harvesting products such as wax or honey from the hive, and these goals can be better met with specific management techniques. All sorts of situations may present themselves in wild colonies that could cause the colony to perish, and many of the issues could be prevented and remedied if that colony were being managed by a knowledgeable beekeeper.

Think about wild dogs and feral chickens—they can survive and even thrive on their own, but you don't have the same investment in their success. When kept as livestock or pets, animals require a different level of care and attention. Honey bees are no different. Plus, plenty of feral animals don't survive when faced with adversity.

Here's an example: in the wild, if a colony goes queenless, the colony will attempt to rear a new queen. However, if that queen can't find enough drones for mating or gets caught in the rain and is unable to return, the colony no longer has the ability to rear a new queen. As all the brood emerge and the worker bees age and perish, the colony will quickly collapse. However, in a managed colony, a beekeeper who regularly inspects their hive will recognize the issue and can provide the colony a frame of worker bee eggs from another colony so they can rear a new queen, give the hive a mated queen purchased from a breeder, or combine the colony with another to at least preserve its resources.

All that said, know that honey bees will never be "domesticated" the same way that dogs or cats become tame and familiar to us as pets. You may treat and think of bees like pets (I know I'm often tempted to do so), but the emotion will never be returned. At the end of the day, honey bees will always be undomesticated animals.

What Is a Hive Inspection, and Why Do We Do Them?

Though each inspection may have different goals, in general I recommend regular inspections to determine:

- The status of the queen bee: Is she alive and well, and is she laying eggs? A colony with a living queen is what we call a *queenright colony*. A colony without a queen is what we call a *queenless colony*.
- Whether the brood is healthy and free from pests and disease.
- Whether the colony has adequate food: nectar, honey, and pollen.
- Whether there is an adequate amount of space in the brood nest and honey storage area.
- Whether you see signs of swarming, which is generally a seasonal concern.

You will learn about each of these topics throughout the remainder of this book.

Before each inspection, I encourage you to think about the "why" of your inspection. Knowing the goals for a hive inspection should guide your actions and can help prevent you from getting overwhelmed with what you are seeing and doing. Writing down the goals for the inspection is even better so you can

Egg Spotting

Egg spotting is one of the most critical skills to learn during your first inspections. The presence of eggs is the *only* way to know for sure whether a queen is laying. Unfortunately, though, egg spotting can be even more challenging than queen spotting for some! Here are my best tips:

1. First, make sure you know what you are looking for. Eggs look like tiny grains of rice adhered to the back of the hexagonal cells in the brood nest. Look around for other, more easy-to-spot stages of brood, like larvae. Usually you'll find eggs in nearby cells.
2. Make sure you keep the sun at your back and tilt the frame so the sun is shining down directly into the cells. Essentially, the sun is the "flashlight" to help you better find the eggs.
3. Some beekeepers find that actual flashlights or magnifying glasses help.
4. If you require glasses to view objects up close, wear them!
5. My best tip is to use your phone as a magnifying tool. This is where a second set of hands or a comb stand can be very useful. Select a spot where you suspect there may be eggs. Finding cells in the brood nest that appear to be empty is a good place to start. Hold a cell phone a few inches away, parallel with the frame, and snap a few pictures. Move to the shade, magnify the image, and voilà! You can instantly see what's in the cell! This is also a good way to see just-hatched larvae, which also tend to be hard to spot.

refer back to them as needed. And after a few years of beekeeping, the written record of hive inspections will be a resource that tells you what to expect in your colonies each season.

When Do We Do Inspections?

Plan to inspect the colonies during the most temperate time of day because this is when an inspection will result in the least amount of interference or potential harm, and when the bees are less inclined to be unhappy with a visit! Extreme cold or heat interferes with the ability of the colony to regulate temperature, and keeping brood out of the hive for long periods during extreme weather can damage the developing bees. In addition, if the foragers are all home because of poor weather conditions, the colony is more likely to be aggressive. Finally, heavy clouds and overcast days can make it more challenging to spot eggs in the hive. Here are a few guidelines for choosing the best weather conditions and times for your hive checks:

- Always inspect a hive after sunrise and before sunset to ensure the foragers are out working for the day. Never inspect a hive at night.
- Choose the mildest part of the day. In the spring months, that may mean avoiding cold mornings and inspecting colonies in the afternoon. In the summer months, this means inspecting colonies first thing in the morning to avoid the warmest temperatures of the day.
- Avoid days with strong winds. A strong gust could blow a queen bee off a frame of bees!
- Unless there is an emergency to address, avoid inspecting colonies when the temperatures are below 60°F. If circumstances require opening a hive, for example if a colony needs emergency feeding, work as quickly as possible and avoid disturbing the brood nest. Damage can be done to brood at these temperatures very quickly.

How Often Do We Do Inspections?

The frequency of regular inspections will depend on the season and your bee-keeping philosophy. Each time a beekeeper inspects a hive, a disturbance occurs. However, regular inspections are important to protect your investment and ensure interventions can be performed in time to keep the colony strong and healthy.

I tend to group inspections into two types: "regular" inspections and "special need" inspections. Regular inspections are important to assess the general health of the colony and to check for the items we previously discussed. Inspections should be done more frequently during the months leading up to and during the time when colonies are bringing in resources, and less frequently during months of extreme cold or heat and when bees will have access to fewer resources. Finding a balance between inspecting hives too much and too little is an important part of your growth as a beekeeper.

Your philosophy will help guide your regular hive inspection rhythm. My personal philosophy and goal is to strike a balance that allows me to minimize interference while also ensuring that so much time does not pass between inspections that I miss an opportunity to provide aid if a colony is weak, low in resources, or struggling with a failing queen.

This philosophy means that I check my hives more frequently during the warmer spring and summer months, and less, or not at all, during the colder winter months. On average, we check our hives once every two to three weeks unless we are monitoring an issue of concern. During the winter months, we do not open a hive at all unless we assess that the colony does not have enough food to survive the winter. If getting into a hive to feed during the winter months is necessary to ensure that the colony survives, move quickly to provide the nutritional resources and take care not to disturb the brood nest.

Of course, special need inspections may be required if you are monitoring an issue in a colony or if an apiary is new and the colonies are just getting established. You will learn more about these circumstances in later chapters.

But First, Light Your Smoker!

A bee smoker must be lit before any hive inspections begin. Lighting a smoker and keeping it lit is not a hard skill to master, but I do find that most new beekeepers fail to spend the time learning this important process in the first year. The problem is that new beekeepers often don't experience the repercussions of not using a smoker effectively, as new, young colonies tend to be small and mild mannered. But by year two, a colony is likely to be much stronger, with many more bees and more resources to protect, and a smoker is necessary to work the hive safely. Unfortunately, a beekeeper who never learned how to properly light and use a smoker may erroneously believe the issue is an overly aggressive colony, when the problem is that they never learned how to manage the bees' natural defense mechanism appropriately. You can find a great video series of how to light a smoker on our Two Hives Honey YouTube channel! Here are the basics:

If you've lit a campfire, you know the basics of lighting a fire. A fire requires tinder, kindling, firewood, and of course oxygen. Some beekeeping stores do sell smoker fuel, but given that the ground is covered with free smoker fuel, there's really no need to buy any!

- **Tinder:** This is your fire starter. Tinder is a material that lights easily but doesn't stay lit very long. My favorite tinder is newspaper.
- **Kindling:** Kindling is material that is less combustible than tinder, but once lit, it burns more slowly than tinder. Examples of kindling include cardboard, small sticks, leaves, straw, pine needles, or charcoal leftover from the last time you lit your smoker. Here, "charcoal" refers to the remnants of the last time you lit your smoker. Do not buy charcoal meant for grilling and use in your bee smoker!
- **Firewood:** This is what will keep a smoker going for a long time. Firewood can be any hardwoods you forage that will fit in your smoker. Oak, mesquite, cedar, pine: any types of wood that you find in your area will do. Choose thicker pieces of wood over small twigs and sticks, as these will stay lit much longer.

- **Oxygen:** The oxygen source is the bellows on the smoker—with each squeeze, the bellows pushes through oxygen to the bottom of the fire.

The key to lighting a smoker is to slowly layer the fuel from the most to least combustible, pumping your bellows as you add more fuel to ensure each layer catches fire. First start with the tinder: make sure it's well lit, and while you pump your bellows add some kindling on top. As you add new layers of kindling, continue to pump the bellows to make sure that each layer catches fire. Once the kindling has caught fire, start adding in the firewood. Be mindful of stuffing the smoker with fuel too quickly—oxygen is necessary as you light the smoker, and adding in too much fuel too quickly can suffocate the fire.

The smoker is well lit when your fire produces thick, white smoke without having to pump the bellows too hard. If you begin to see flames shooting over the top of your smoker when you work the bellows, close the lid for a moment so that the fire gets less oxygen and lessens the fire a bit—you don't want a stray spark to start a fire elsewhere.

Over time, the fuel in the smoker will need to be replenished. The smoker is running low on fuel when ashes spray from the smoker as you squeeze the bellows. When this happens, open up the smoker (use a hive tool to pry it open—

the smoker will be hot!), add more firewood, and pump the bellows to stoke the flames once again.

Practice lighting the smoker *before* the bees arrive! Having already mastered lighting a smoker will mean one less concern during hive inspections.

How Do We Perform a Hive Inspection?

Before beginning any inspection, think clearly about the goal of the inspection. Setting and reviewing a plan will ensure that you are focused on the task at hand and that you are prepared with all the tools and equipment you need to be successful. This helps you meet your goals during the inspection, and it hopefully means avoiding getting halfway through the inspection and realizing you forgot something, such as an additional hive body to allow the colony more space to grow.

Before starting, make sure to:

- Check the last inspection notes and review them in conjunction with the goals for the day. If the last inspection found a weak colony or a colony low on honey, you may need to be prepared to feed or share resources with this hive.
- Think about where you are in the bee season. Is this a time of year when bees usually have access to lots of nectar and can make honey? If so, be prepared with extra space for the colony to store honey: namely, adding extra hive bodies and frames to a Langstroth hive or moving the follower board in a top bar hive so the bees have access to more top bars. If this is a time of year when the brood nest will begin contracting soon (for example, as the colony moves into winter), you should be prepared to consolidate the hive. You will learn more about the beekeeper's calendar and about seasonal action items in chapter 12.
- Have your note-taking tools ready. Note-taking is important, even for experienced beekeepers. Our memories are *not* as good as we think they are! For example, during the last hive inspection, did you have trouble finding eggs? This is an important item to note so during the next inspection you are sure to look for signs of a laying queen.
- Suit up! Being concerned about getting stung because you aren't wearing the proper protective gear will be an unnecessary distraction.
- Light a bee smoker, and light it well, every time a hive is opened. No exceptions!
- Have a hive tool at the ready. Honey bees will use propolis to glue all of the

hive components together, so a hive tool is a must-have for inspections. I always recommend having two on hand in case one gets misplaced.

Now that you've got your tools together and you've reviewed the plan for the day, you are ready to inspect the hive! Please note that the following instructions are for inspecting a Langstroth hive. If you are using a top bar hive, the general steps are the same but you won't be removing boxes. Instead, you will be removing your outer cover and then pulling out top bars to look inside.

1. Take a moment to observe the hive entrances. Note which hive has the most bees coming and going. Usually, the hive with the most activity is the strongest colony and the hive with the least activity is the weakest colony. Work the weakest colonies first, and work through to the strongest colonies. This allows you to determine whether the weaker hives could use a boost. You can give them this boost by sharing resources from the stronger with the weaker colony. Also, if there is any unusually busy activity at any hive, and if the bees appear to be fighting one another, the colony may be getting robbed. This is when a stronger colony invades a weaker one and steals its honey or nectar. We'll talk more about why this happens in chapter 7.

2. Once you've determined the order of operations for working the hives, give all the hives a few puffs of smoke at the entrances. Smoking all hives in close proximity will help get ahead of any aggressive tendencies. Smoke the entrance of each hive once again before you open that hive's outer cover, and continue to apply a few puffs of smoke as you remove each piece of the hive.

3. Using a hive tool, pry the outer cover off the hive. If you have Langstroth hives, turn the outer cover upside down on the ground or on a work table, if you have one. This will act as the "stand" for any hive bodies pulled off the top of the hive. Do not put hive bodies directly on the ground, for two reasons. First, bees will run from the light, and they (or your queen!) could end up on the ground. Second, the bottom of the hive body will be covered in sticky propolis, and if you place it directly on the ground, grass, rocks, sticks and other debris will stick to the bottom of the box.

4. If you have a Langstroth hive, use a hive tool to remove the inner cover and then apply a few puffs of smoke inside the hive. Keep the inner cover nearby. Use this to cover whichever stack of hive bodies you aren't currently inspecting. This helps with temperature control and keeps the bees a bit more relaxed during an inspection.

5. Once inside the hive, the first order of business is to inspect the brood and confirm whether there is a laying queen. This means first finding and

inspecting the brood box. The brood should be located in one or both of the bottom hive bodies. In a top bar, the brood is generally located at the front of the hive, closest to the entrance. A Langstroth hive may have honey supers on top of the hive that need to be removed, or you may find yourself in the brood nest as soon as you remove the inner cover.

6. If you do have to remove a box or two to get into the brood nest, use your hive tool to break the propolis seals between the boxes. Then, as you lift one end of the box up, apply a twisting motion to help break the remainder of the propolis seal. Rest your hive body on your upturned outer cover. Resting the hive body kitty-corner to the outer cover limits the number of touch points between the equipment and limits the number of bees you may squish in the process. Remember, never put hive bodies directly on the ground. Having something underneath them (a table, a piece of cardboard, or preferably your inverted outer cover) keeps your bees and queen from running out of the hive box and into the grass.

7. Remember to apply smoke as you remove each piece of your hive!

8. Once in the brood nest, the area where the eggs, larvae, and pupae are found, use a hive tool to pry out the first frame or top bar. I like to use the hive tool to break the propolis seals on all four sides before I pull a frame or top bar. Propolis can be incredibly strong, and this will help protect your frames from breaking as you remove them.

9. Once you remove a frame or top bar, try to keep the frames as perpendicular to the ground as possible. Meaning, don't hold the frames as you would hold a plate of pancakes! Holding frames, and especially top bars, parallel to the ground can cause them to fall apart under the weight of the honey or brood. When you are ready to place the frame back into the hive, be sure to use a hive tool to clear enough space and then slowly push the frame back into place. Move slowly so the bees have time to move out of the way and prevent being squished!

10. Remove as many frames or top bars as necessary to complete the goals of your inspection. I like to have one to two frames removed from the hive body at all times to allow myself some space to work. A comb stand is a great tool to safely hold these resting frames out of the hive. Remember, in every regular inspection you should answer the following questions:

- What is the status of the queen bee? The presence of worker bee eggs report that she is alive and laying!
- Is the brood healthy and free from pests and disease? You will learn more about how to spot pests and disease in chapter 10.
- Does the colony have access to proper nutrition?

- Is there an adequate amount of space in the brood nest and honey storage area?
- Are there signs of swarming? You will learn more about swarming in chapter 8.

Look for these in each regular inspection, but you may also want to monitor other issues of concern. For example, if a colony went queenless, you will assess whether they have requeened themselves.

11. Once your inspection is complete and you begin to rebuild the hive, use smoke or a bee brush to move bees out of the way to avoid any unnecessary bee squishing. Place each hive body at a slight angle to avoid too many contact points. Once contact has been made, slowly shift the box back into place, gently moving the bees out of the way as you work.

Common Colony Conundrums

A WEAK COLONY

The strength of a colony depends on the number of worker bees it has to perform all the duties necessary to keep a hive safe, clean, and fed. Therefore, a weak colony may have too few bees to perform all the necessary functions. One of the concerns with weak colonies is they aren't able to adequately guard the hive from pests and predators. Simply adding or taking away space in a hive can create a weaker or stronger colony. This is why, during inspections, beekeepers should always evaluate whether a colony needs more, or less, space.

If a colony is weak, there are several ways to help protect its resources. First, consolidate the hive so that the bees present can adequately guard the space. Also, consider narrowing the entrance of the hive with an entrance reducer for the same reason.

Another way to help a weak colony is simply giving it brood resources from another colony—that is, pulling a frame of brood from a strong colony and placing it in a weaker colony. Try to become accustomed to thinking of the apiary as a whole unit, and remember that all sorts of resources can be shared between hives very easily to help boost the populations of colonies that are weak. When sharing brood resources, shake off most of the bees on the frame to be shared. If a colony does not have an adequate number of adult worker bees, try to share a frame of capped brood. Capped brood requires the least amount of feeding and care by the nurse bees and is closer to emerging and providing new bees than a frame of larvae or eggs. If a queenless colony needs the opportunity to rear a new queen bee, share worker bee eggs. Above all, make sure you are taking resources from a colony strong enough to spare the brood.

Smoke 'em if You've Got 'em

A bee smoker is a ubiquitous image when beekeeping comes to mind. But what does it do, and why won't I stop talking about its importance? In short, a smoker interferes with the communication system of a colony.

Let's first review what you have learned about how a colony defends itself. Pheromones are critical in a colony's defense system. After a worker bee stings a predator, she flies away, leaving her stinger behind in the skin of the unfortunate animal or human. In the process, her stinger rips out her insides and releases the alarm pheromone, warning her sisters of the danger. Furthermore, the bees at the entrance of a hive, the defender bees, can release this pheromone if they feel the colony is being attacked, prompting other bees to defend the colony.

The smoker interferes with the receipt of alarm pheromones by the rest of the colony. The message may be sent, but it's masked by the smoke's strong aroma. The bees are none the wiser, and so beekeepers can work safely, protecting not only the beekeeper but also the bees themselves since the act of stinging will end a honey bee's life. There's not a lot of "magic" in a bee smoker. My bee smoker on any given day will contain leaves, pieces of mesquite and oak wood, and probably a lot of cardboard shipping boxes.

Applying a few puffs of smoke at the entrance of the hive will mask the pheromone of the defender bees, and smoke also can mask the pheromone if and when you take a sting or two. I often work gloveless, and a shrewd observer will note that whenever I take a sting to the hand, I always hold my hand in the path of the smoke for a few brief moments, masking the alarm pheromone.

Be careful with jumping to feed a weak colony. There is a tendency to want to feed weak or sickly creatures, and perhaps nutrition is an issue, but feeding won't fix all problems. See chapter 7 to learn more about feeding bees.

Finally, try to determine the reason for the weakened colony: perhaps the brood is diseased or a queen bee is failing. Requeening a colony can help solve a lot of issues.

A LIGHT COLONY

A light colony is defined as one that does not have enough stored honey and nectar resources to feed its young and/or the adult bees in a hive. We use the term "light" because honey and nectar are significantly heavier than brood frames. An experienced beekeeper can perform the "lift test": picking up the back of the hive to estimate how much food a colony has stored. If a colony is light on nectar or honey, you can share food from another hive. Often beekeepers will choose to freeze a few frames of honey during a harvest, and these can be fed back to a colony later if necessary. Just make sure to defrost the frames first. A colony can also be fed sugar water if none is available from other colonies to share. Never feed a colony store-bought honey. Most store-bought honey is infected with brood diseases that don't affect humans but can cause great harm to a colony. If a colony is low on pollen or bee bread, the food necessary for brood rearing, you can also share these between hives or feed it store-bought pollen substitutes meant for feeding bees that can be found at beekeeping supply stores.

A QUEENLESS COLONY

A colony must have a laying queen the majority of the year to ensure that the population stays strong enough to perform the activities keeping the colony alive and well. Examine your colony on a regular basis to determine the status of the queen by looking for worker bee eggs. You will learn more about queens in chapter 6, but there are several steps to help a queenless colony. You can let the colony requeen itself or buy a queen from a breeder. If the colony has tried to requeen itself and failed, you must intervene with a breeder queen or a frame of eggs to let them try again, or the colony will fail. Be sure to share frames of capped brood from another strong hive to keep the population steady while the colony works to requeen itself.

A Few Best Practices

Hive inspections can be overwhelming. There is so much to learn, and much to observe and note when inspecting a hive. Here are a few best practices:

1. It cannot be said enough times: always light your smoker. If you are having trouble getting and keeping it lit, practice on days when you won't be inspecting a colony. One bad experience is enough to scare a beekeeper away from beekeeping forever, and most of these experiences can be avoided by using a smoker properly.

2. Cover your face! Some beekeepers, including myself, choose to wear less gear. But I always have my face covered, and I have gloves and a full suit within arm's reach at every inspection. No medals are handed out for wearing less gear—stay safe, and only make the move to use less gear when you feel you have mastered the art of the inspection.

3. Keep notes! Your learning will be expedited if you can look back on how the colony changed over the course of a year, and it will help inform seasonal decisions in years to come.

4. Not every inspection necessarily entails pulling every frame. It's rare that I find the need to inspect every frame in a hive. If I do, it's normally because I have to find the queen or I am looking to consolidate a hive in some way. Inspections should be more involved initially, as you practice identification and become more familiar with what's going on in the colony. But as your experience grows, pulling just a frame or two may tell you all you need to know about a colony and its health.

5. The frames and top bars do not have to be returned to their original location. However, the brood nest should always be left intact. This means any frames with brood should be kept together in the hive. In my Langstroth hives, I always make sure my brood nest is in the lower boxes, centered in the middle of the boxes with a frame of honey around the outer edge of the nest. Honey is a great insulator.

6. Move slowly and try not to panic! Moving methodically and slowly kills fewer bees and disturbs the hive less than moving quickly and throwing equipment around. Moving in this manner also leads to less chances of getting stung. Don't fall into the trap of moving so quickly that you forget to use the smoker. Taking the extra few seconds to do so will pay off in spades. If you get overwhelmed and need a breather or a drink of water, put the hive boxes back together and lay the inner cover over the top, step back, and compose yourself.

7. Respect the bee space. Bee space is the gap left in between drawn sheets of comb that enables bees to pass each other and work freely in the nest. Colonies, either managed or wild, will always follow the law of bee space. There is a range of distances bees use for bee space, but generally the distance most often cited is ⅜ inch. This means bees will leave ⅜ inch

Lessons from a Beekeeper

At some point in your beekeeping journey, you will get stung. I did not take my first sting until after a year as a hobbyist beekeeper, and the anticipation of what it would feel like was worse than the actual sting. (I actually had already quit my job to start Two Hives Honey before I took my first sting. Good thing I didn't learn I was allergic, huh? First lesson: confirm you don't have any medical conditions that would ruin your plans to start a business *before* you quit your job!) Every beekeeper has her least favorite place to get stung: for me the most painful spots are on an eyelid, inside my nose, and on my upper thighs. But that is just anecdotal: I always look for scientific proof to back claims. Good thing a graduate student at Cornell University decided to sacrifice his body in the name of science to determine the age-old question: Where is the most painful place to get stung on the body? The student made himself a literal human pin cushion, taking 5 stings a day over 38 days, for a total of 190 stings. He chose twenty-five different sting sites, from his head all the way down to his middle toe.

The results of his research supports my hypothesis, as he found the nostril was the most painful sting site, followed by the upper lip, penis shaft, and then scrotum. Upper thigh didn't make the top ten, coming in at nineteenth on the list.* We may just have to agree to disagree on this one. To be fair, though, I have never been stung on the penis shaft or scrotum.

* Michael L. Smith, "Honey Bee Sting Pain Index by Body Location," *PeerJ*, no. 2 (2014): e338, https://doi. org/10.7717/peerj.338.

between drawn combs, which is just enough space for two bees to walk and work back-to-back. They will leave about half that amount of space around the outer edges of the hive, where bees have only one side of one sheet of comb to work and need space for only one bee. Langstroth hives are designed to mimic the bee space bees will leave in the wild. Be sure to respect the bee space by never leaving more than ⅜ inch of empty space in a hive. All frames should get put back into every hive body. If a frame is left out, and the bees get on a nectar flow, they will draw comb in any space larger than ⅜ inch and there will be quite the mess next time you try to inspect they colony. Similarly, never leave an empty hive body on a hive with no frames! Be sure the frames are evenly spaced across the hive body and not shoved to one side or another. If you are using a top bar hive, make sure the top bars are pushed very tightly together—you should see no space between the top bars.

8. Do not give bees more space than they can adequately guard! Keep in mind a colony is like an accordion, expanding and contracting throughout the year in response to the availability of resources. Manage your space accordingly.

9. Keep colonies strong and equalize the resources often. Don't let one colony slowly fail while another colony is bursting at the seams. Sharing resources can not only help weak colonies but also prevent strong colonies from swarming, as you will learn in later chapters.

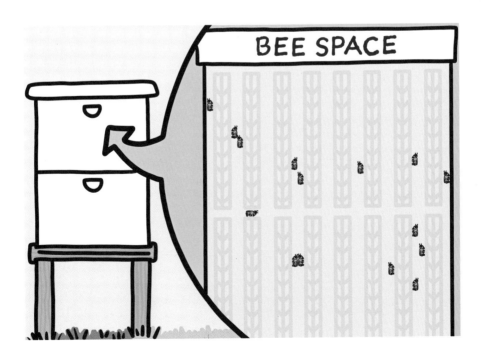

CHAPTER 6

Understanding the Queen Bee

By now, you've probably realized how critical a queen's role is to a colony. She is the only bee whose job it is to produce offspring, and she is the connecting glue in a colony. Because she is the mother of all the bees in the colony, her genetics are important: they will help determine a colony's resistance to disease and its temperament, just as your own family's genetics can determine your predisposition to certain health conditions.

We've learned the key to a queen's success is her ability to produce pheromones, the most important of which is called the queen mandibular pheromone (QMP). The QMP is secreted by glands above the mandibles, or jaws, of the queen bee, and all the bees are attracted to this pheromone. Through grooming her, the bees spread her pheromones throughout the hive. This is how her retinue, the eight to ten bees whose job it is to feed her and clean her, know how to find her. The queen's pheromones are also responsible for keeping a swarming bee cluster together. If the queen dies or is unable to fly, and therefore the pheromones are lost, the cluster will return to the hive. They are also how drones find virgin queens for mating. You will learn in chapter 8 about how a congested hive can cause these pheromones to wane, alerting a colony that it's time to reproduce through swarming.

And finally, these pheromones are very important because they communicate her physical well-being, or lack thereof, to the colony. If the pheromones wane because she either is sick or has died, the absence of the pheromones stimulates workers to rear a new queen.

Queen Spotting

Beekeepers tend to put a lot of emphasis on queen spotting, and I'll admit, it is a fun, albeit sometimes frustrating, task. However, the occasions a bee-keeper will need to spot a queen are usually rare. Rather, what's more critical is learning to spot eggs, as that is the only way to know for sure if a queen is actually laying offspring. That said, the time will come when it's critical for you to spot a queen, and here are some tips:

1. Know what to look for! Know the attributes that distinguish queens, workers, and drones. Unlike workers and drones, queens have abdomens that extend beyond the end of their wings and no hair on their thorax. Because drones are so large, they are commonly mistaken for queens. If you believe you've found a queen, look at the eyes: if they are larger than those of the worker bees and centered toward the top of the head, and if the wings are even with the abdomen, it's likely a drone.

2. Look in the brood nest first. Your queen can be found in any area of the hive available to her, but a best bet is to look for her where she per-forms her duties—the nest.

3. If finding your queen is a must and you have Lang-stroth hives, have a second hive body nearby to hold the frames that you "clear" as you dig through the box. Once you have confirmed that she is not on a frame, put the frame of bees into the extra box.

4. Queens are going to run from the light, so another trick is to clear one or two frames in the hive to give you some space to work. (A second set of hands is helpful for this task.) Next, push all the remaining frames together in pairs of two, with light shining through on either side of each pair. Wait a few minutes, then carefully remove one of the pairs of frames, holding them together. Open the frames like a book, and quickly run your eyes over the sides of the frames that were previously facing one another. There's a good chance you will find a queen in the darkness created by the pairs of frames.

A Queen Is Born

A queen bee can lay two types of eggs: fertilized eggs, which produce female worker bees, and unfertilized eggs, which produce drone bees. So how do we get queen bees?

There are no queen bee eggs; rather, queen bees are created from worker bee larvae. The simple explanation is that queen bees are created when worker bees feed a special diet to worker bee larvae. All larvae are fed a diet of royal jelly for the first three days after they hatch, but if a colony has reason to produce a queen bee, perhaps because their queen has failed or is failing, or the colony is preparing to swarm, they will begin feeding extra royal jelly to one or more worker bee larvae just after it hatches. They will also build a longer cell around the developing queen bee, which is necessary because her body is so much larger than that of a worker bee. This queen cell, as it is called, is shaped like a peanut. At the point that worker and drone larvae are switched to a diet of bee bread and nectar, worker bees will continue feeding a queen bee a diet of royal jelly.

It's important to know that queen bees have the same development cycle as a worker bee: egg, larva, pupa, then adult. But the timeline to rear a queen is slightly shorter. The egg will hatch on day 3 ½, but the larva grows more quickly and is capped around day 8, and the adult queen will emerge on day 16. Understanding this process and how long the queen will spend at each stage will help guide you to understand how long it has been since your colony went queenless and how long it will be before your hive has an emerging queen. You've already seen this illustration, but because understanding this is so important, I've included the development cycle graphic for all three castes once again.

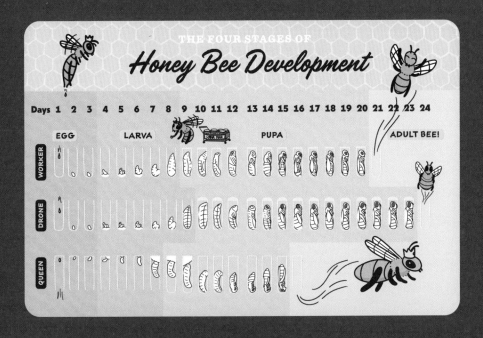

THE FOUR STAGES OF

Honey Bee Development

| Days | 1 | 2 | 3 | 4 | 5 | 6 | 7 | 8 | 9 | 10 | 11 | 12 | 13 | 14 | 15 | 16 | 17 | 18 | 19 | 20 | 21 | 22 | 23 | 24 |

EGG   LARVA   PUPA   ADULT BEE!

WORKER

DRONE

QUEEN

Once the first virgin queen emerges from her cell, she will move around the hive and use her stinger to kill the other developing queen bees through the walls of the queen cells. If two queens emerge around the same time, they will fight each other until one is victorious. This virgin queen must now leave the hive to mate with drones.

Long Live the Queen!

Queens have a life expectancy of two to three years, and their life span is dependent on their ability to lay fertilized eggs. As she ages or runs out of sperm and can no longer produce as many workers as the colony needs to survive, the workers will take action. Without a viable laying queen, a colony will quickly perish. The workers will work quickly to replace her, rearing another queen from one of the queen's daughters. Then the workers will form what beekeepers call a murder ball and ball around the queen tightly, suffocating her. Of course, other events can lead to a queen's demise, such as a beekeeper accidentally squishing her. (If this happens, know that at some point most beekeepers who keep at this long enough accidentally kill a queen. It's heartbreaking, but you aren't alone. Keep reading to discover my own queen murder mystery!)

Performing regular inspections outside of the winter months and looking for eggs during each inspection is critical to knowing the status of your queen. If you do not find eggs in an inspection, your colony may be queenless. When this happens, you can let the colony requeen itself or buy a mated queen from a bee breeder for your colony. However, the timing of when you realize your colony has gone queenless will dictate the available options. If a colony is given a new queen and has already produced a queen of their own, they will not accept another queen bee!

If a colony goes queenless and they have viable worker bee eggs when they lose their queen, the bees will begin rearing new queens in queen cells. They may begin rearing just a few or more than a dozen queens. Keep in mind that it takes sixteen days for a new queen to emerge, so if a colony went queenless more than sixteen days ago, you may be too late to see the intact queen cells. Instead, look for evidence that your colony has requeened itself: you should see one queen cell from which a queen emerged, and the rest of the queen cells will be ripped down, either by the new queen herself or by the workers after she killed the developing sister queen bee inside.

If a colony has gone queenless and a virgin queen has already emerged, you must give the colony the opportunity to complete this biological process before trying to give them a queen procured from a breeder. Once the colony goes queenless, the total time it takes for a queen to emerge, leave the hive to mate,

return, and begin to lay eggs is around thirty days. After thirty days you should see new, fresh eggs. And remember, you will have to wait an additional twenty-one days for new emerging worker bees, so your colony will continue to shrink in size until this process is complete.

If you don't see new eggs after thirty days from the time the colony went queenless, likely something happened to the virgin queen on her mating flight. Perhaps she got lost, got caught in the rain, or was eaten by a predator. At this stage you can give your colony a frame of eggs from another hive to try to rear another queen, or you can give it a mated queen from a breeder. If you choose the first route, keep in mind that at this point a colony may be severely weakened from going so long without a queen producing new bees. You may have to continue sharing brood resources from other hives to ensure that it is strong enough to try to rear another queen. In most cases of a colony failing to produce a queen on the first try, I recommend getting a queen from a breeder. If this is not an option or you are worried the colony is now too weak to continue, you can combine this colony with another in the apiary. You will learn about combining colonies in chapter 9.

If you catch your colony in the act of rearing a new queen, or they are not successful at producing a new queen that mates and returns to the hive, you can install a queen from a breeder. First, cut down all existing queen cells that may be in the hive with a hive tool. If any queen cells remain, a colony will not accept a new queen. Next, slowly introduce the queen via the queen cage. Releasing a queen into a colony too quickly, even if the colony is queenless, will result in her demise! The bees must become accustomed to her pheromones first. Breeders

Lessons from a Beekeeper

Never underestimate the power and ability of a honey bee colony. Humans tend to minimize the brainpower and abilities of creatures smaller than us, and even though a bee's brain is less than two cubic millimeters in volume and 0.0002 percent the size of the human brain, honey bees demonstrate incredible intellect and comprehension. What they lack in individual physical strength they make up for in sheer intelligence. If a mammal, such as a mouse, enters the hive to try to build a nest or steal some grub, the bees can sting the animal to death but cannot remove the carcass. Instead, they will coat the entire body in propolis to ensure the rotting carcass does not "contaminate" the hive. Some of these abilities can't easily be explained. I was once working hives with students and we made the fatal mistake of pouring sugar syrup into a feeder without checking it for bees first. As we poured the syrup into the feeder in the hive, the queen came floating to the surface. I tried to save her, but my actions made the situation worse and she fell to the bottom of the feeder. I quickly poured the syrup out on the ground and gently picked her up. Bees breathe through openings along their abdomen called spiracles, so when these openings become clogged, they suffocate. I watched as worker bees quickly surrounded her as she writhed and eventually lay still. She did not move for several minutes, and I knew that I had unintentionally killed her. I placed her gently in the hive so that the bees could recognize her death immediately and begin rearing a new queen. I've done all sorts of damaging things and caused unintentional harm when working bees, but the guilt about this one was some of the hardest to bear. I returned a week later expecting to find queen cells, and instead I found a healthy queen laying eggs, just as she had before. It was far too soon for the colony to have requeened itself. Instead, the bees had been able to resuscitate the queen bee that I was certain I had watched die in front of my eyes.

use all sorts of different queen cages for this purpose. Most have a fondant candy plug, which will prevent the workers from releasing the queen bee prematurely before they can accept her as their own. Check with the breeder for specific instructions and advice on how to introduce the queen to your colony safely.

Requeening a Colony

There may come a time when you find reason to proactively replace a queen in a colony. If this is the case, make sure to find and remove (or kill) the original queen first. A colony that is queenright, meaning it has a living queen, will not accept your new queen. Waiting twenty-four to forty-eight hours after the queen is removed before introducing a new queen may increase the chances of your colony accepting your new breeder queen. If you choose to wait, be sure to cut down all developing queen cells before installing the new queen.

Reasons to requeen a colony may include:

- A colony is demonstrating a level of aggression that is unacceptable or unsafe for you and your family. The temperament of a colony comes from the genetics of the queen mother, so replacing her can mean new offspring that don't react as aggressively. In deciding whether a colony is overly aggressive, be sure that you are lighting your smoker for every inspection and using it effectively. If you don't, and you simply aren't effectively masking the alarm pheromones, requeening will not solve the issue. Additionally, assess whether the aggression is occurring at every inspection and getting worse over time. If the aggression is actually an isolated event because of a predator incident or a beekeeper inspecting a hive on a bad weather day, requeening may not be the answer.
- A queen seems to be failing, laying in a sporadic and spotty pattern or not laying prolifically even when she has ample space, food resources, and nurse bees to care for the young.
- A colony is demonstrating recurring symptoms of disease, viruses, or mites.

Some beekeepers requeen their colonies every one to two years. This is not a practice I keep in my own apiaries, as I simply requeen as needed. Let's examine the arguments for and against a proactive requeening approach.

Arguments for a proactive and recurring requeening schedule include:

- Requeening on a schedule can help always ensure a young, well-mated queen in a hive. Of course, a colony can go queenless in an off year or in between queen replacements. But this approach can help ensure that a queen doesn't fail when you aren't looking. This approach creates what we call a forever-young colony.

- If you kill or remove a queen and let the colony go broodless for a bit before installing a new queen, the colony will experience what is called a brood break: a period when there is no brood in your hive. Periodic brood breaks are good for the health of the colonies, because they can help clear up any disease of the brood and also help control for varroa mites. You will learn more about brood breaks and varroa in chapter 10.

Arguments for following a more reactive queen replacement strategy include:

- Queens aren't free, and they actually can be quite expensive. Also, you may not always have access to a breeder queen.
- Requeening a colony isn't always a successful endeavor. The colony may not accept your new queen, or your breeder queen may be a dud.
- Philosophically, it may just not sit right with you. Knocking off a perfectly fine and healthy queen isn't my favorite thing to do, and because of that I act more reactively.
- You have to find the original queen. This is a difficult task for many beekeepers.
- Not requeening each year may make you a better beekeeper, as you have to be more in tune with what's happening in your hive. Of course, this is not a given, but to better ensure success you do have to pay more attention to her brood pattern and the state of the brood.

Worker Bees Can't Lay Eggs . . . Can They?

The queen is critical to a colony because she is the only one that can produce offspring . . . with one exception!

Worker bees actually do have ovaries, which are suppressed from full development by a combination of the QMP and pheromones emitted from eggs and larvae in a colony (open brood). If a colony loses its queen and the workers are unable to successfully rear a queen that returns to the hive mated, the colony does not have the ability to try to rear another queen.

(Unless, of course, they have the good fortune of being cared for by an observant and educated beekeeper who gives them a frame of eggs to try again!) Over time, all the open brood will become capped by the nurse bees and then will emerge as adult bees. Eventually, no brood will remain. A colony cannot survive without the workers responsible for all operations of the hive, so the colony is doomed to perish. In a colony without a live queen or open brood, the pheromones that keep workers' ovaries from developing dissipate, and soon the worker bees begin to lay eggs. Because worker bees cannot biologically mate, they can lay only unfertilized eggs, producing drone bees. This is what we call a laying-worker colony. The evolutionary reason for this phenomenon is that when a colony is doomed to fail, this is a last-ditch effort for a colony to share its genetics. The hope is that one or more of these drones will be able to mate with a virgin queen, thus ensuring that the genetics of the colony live on. The signs of a laying-worker colony include:

1. Multiple eggs per cell. Workers are not adept at laying eggs, and you may see anywhere from a few to many eggs in each cell.
2. Eggs off center in the cells. Workers' abdomens are much shorter than queens', and they can't reach the back of the cell as easily as a queen bee can. Plus, their aim is pretty poor! This means eggs will be off center or along the side wall of the cell.
3. Drones developing and emerging in worker bee cells. Unlike queen bees, workers won't lay their drone eggs only in the larger drone cells. This means that in a laying-worker colony, drones will be emerging from regular worker bee cells too. As the drones develop in worker cells, you may see that the workers have altered these cells so that they stick out a little from the frame to effectively house the larger drone larvae.

A colony that has gone queenless for six weeks or more and is exhibiting the signs above is probably a laying worker hive.

What can you do about a laying-worker colony? Laying-worker hives are doomed to fail without intervention, and in my experience the work and resources necessary to save a laying-worker colony probably aren't worth the trouble. The workers' ovaries must first be suppressed before providing a new queen, which is done by providing a frame of open brood from another colony every week for three weeks. Only then will a laying-worker colony accept a breeder queen or begin to rear a new queen from eggs. The amount of time and resources this requires, all while the colony continues to shrink, makes this a challenging endeavor. Rather, I shake all the bees out of the hive and take the hive equipment out of the yard. A number of the bees will disperse to other hives, ensuring that no one hive ends up with all the laying workers from the colony.

CHAPTER 7

Feeding Honey Bees

"Should I be feeding my bees?"

This is one of the most commonly asked questions by new and inexperienced beekeepers, and unfortunately far too many folks out there are ready to jump in too quickly and advise. I find a lot of "internet beekeeping" advice very frustrating, but this is probably at the top of my list. I always advise my students that if someone jumps in and starts answering this question without first asking questions about the hive, stop listening immediately!

The question of whether to feed bees does not have a one-size-fits-all answer. Whether you should be feeding, how much, and in what ratios will vary based on the strength of the colony, where you are in the bee season, when the next nectar flow is expected, your personal goals and philosophy, and, of course, what nutritional resources your bees have stored in the hive. In order to advise a beekeeper in good faith about whether they should be feeding their bees, I must know a lot more about the context of the colonies.

If you are preparing to feed your bees, ask yourself, "What is my goal in feeding?" Although the primary reason for feeding bees is to remedy nutritional deficiencies in the colony, other reasons beekeepers choose to feed include to encourage comb building, to stimulate brood rearing, to aid in queen rearing, or to help clear up certain brood diseases. Next, note the current state of the colony. For example, if the goal of feeding is to address nutritional deficiencies, understanding the strength of the colony helps determine how much feed to provide and at what intervals.

Honey bees ideally can forage for all the necessary nutrients they need from the environment, but most beekeepers, at some point, will need to feed their bees. When beekeepers talk about feeding bees, they are usually referring to feeding them a syrup made of sugar water to replace carbohydrates found in nectar, but it also may include feeding pollen substitutes to replace the protein and fats found in pollen. This chapter will discuss what, how much, and when to feed.

Remember that understanding the nutritional needs of a colony and knowing how and when bees can access these nutrients from the environment helps guide decisions about whether feeding is necessary. Because the primary reason for feeding is to replace the macronutrients found in nectar and pollen, it is critical to understand why these nutrients are necessary, recognize when the colony is deficient, and know the consequences of this deficiency. I recommend reviewing chapter 3, "Honey Bee Nutrition," to refresh your memory about the nutritional needs of bees throughout their development cycle and their importance in supporting colony processes to help provide context for this chapter.

What to Feed Honey Bees

As nuanced and complicated the topic of feeding bees can be, the answer of what to feed is actually quite simple.

If you are feeding to replace carbohydrates (nectar and honey) choose a syrup made of water and refined white table sugar. The ratio of sugar to water may vary slightly depending on the goal—whether to prevent spring starvation of a colony, to ensure adequate winter stores, or to stimulate comb building. If stimulating or supporting brood rearing, you can use sugar syrup and pollen substitutes. We will discuss pollen substitutes later in the chapter.

Plain white table sugar is the closest substitute for nectar available to beekeepers. The composition of nectar varies from plant to plant, but plants pollinated by bees tend to be sucrose-dominant. Similarly, table sugar is 99.95 percent sucrose.

What Not to Feed Honey Bees

The list of what not to feed honey bees is as long as the list of what to feed your bees is short. These include:

- Light and dark corn syrups
- Agave
- Brown sugar
- Organic sugar
- Turbinado sugar
- Powdered sugar
- Molasses
- High-fructose corn syrups
- Organic cane sugar

Many of these sweeteners contain additives that are bad for bees. For example, many of the corn syrups used in pie making contain vanilla and salt.

High-fructose corn syrup is typically used by commercial beekeepers because of its long shelf life, and it isn't in itself bad for bees. However, when heated to higher temperatures (around 113°F) it produces hydroxymethylfurfural (HMF), which is toxic to honey bees. The way the syrup typically is stored and transported means there's a high likelihood that it reaches these temperatures at some point in its handling.[1]

Many are surprised to learn that organic cane sugar is on the "do not feed" list. It is easy to anthropomorphize bees and associate what we see as healthy for humans as also healthy for bees. But this logic is misguided. The extra processing involved in refined table sugar includes a bleaching process that removes some solids that would otherwise contribute to dysentery in bees.

Finally, and most importantly, do not feed honey bees honey purchased from the grocery store. First, honey at the grocery store may not even be real honey, as, unfortunately, adulterated honey is very common. Adulterated honey is "honey" that has been cut with cheap sweeteners like high-fructose corn syrup and is not truly 100 percent honey. Second, grocery store honey is usually blended honey, which means honey from many beekeepers, regions, and even countries has been mixed together and bottled. Any spores of different brood diseases can spread very quickly during the blending of this honey. If fed to honey bees, these spores can infect your colonies as well. Feed honey bees only honey whose origins you can trace: honey produced by your own colonies.

MAKING SUGAR SYRUP

The most common ratio recommended for feeding honey bees is one part sugar to one part water. This ratio can be measured by volume or weight; they are close enough that either is acceptable. Measuring and making sugar water is one of those activities that seems to give beekeepers great consternation. Don't worry! If your syrup is slightly heavier, meaning with a higher concentration of sugar to water, or slightly lighter, meaning with a higher concentration of water to sugar, it's OK. In this case, close enough is good enough. Lightly heating the water before adding the sugar can help dissolve the sugar. However, unless you are making a heavier syrup, hot water is not necessary. We usually make our 1:1 ratio of syrup right in the bee yards. If you do decide to use heat, do not boil the syrup. Boiling the water only before adding the sugar can make the syrup last a bit longer before fermenting, but boiling the sugar water produces HMF, which you've learned is bad for bees.

Two situations may arise that require a different ratio than the 1:1 sugar to water. First, if you are feeding a colony in the late summer and fall because of a

RECIPE

1:1 Sugar Syrup

YIELD: ~1 GALLON

INGREDIENTS
5 POUNDS SUGAR
10 CUPS WATER

DIRECTIONS
Mix water and sugar. Using warm water can help the sugar dissolve faster.

RECIPE

2:1 Sugar Syrup

YIELD: ~1½ GALLONS

INGREDIENTS
10 POUNDS SUGAR
10 CUPS WATER

DIRECTIONS
Heat the water before adding the sugar, then mix.

shortage of stored honey, I recommend feeding a heavier syrup with a minimum of two parts sugar to one part water. A heavier syrup provides more carbohydrates to the colony faster, which is important if winter is rapidly approaching. Second, a heavier syrup tends to encourage bees to store the syrup in the colony, which is important if the colony is short on winter honey stores. Also, heavy feeding of lighter syrups late in the year can result in excess water stored in the hive, which can later result in condensation in the colony. Condensation is very dangerous for a colony in the winter months. Note that heavier syrups will require warmer water to adequately dissolve all the sugar.

A second exception to the 1:1 ratio is if a colony requires feeding during the coldest winter months. If temperatures are regularly below 50°F, the honey bees are spending their days tightly clustered to generate enough heat to keep the queen and cluster of worker bees alive. Clustered honey bees cannot break their cluster to seek out sugar syrup from feeders.

Hopefully, any nutritional deficiencies can be overcome with late summer and fall feeding of heavier syrup. But if a colony is still potentially too light to survive winter, I recommend feeding dry sugar through a method called *mountain camp feeding*. In this method, simply lay a single sheet of newspaper across the top of the frames in your Langstroth hive and carefully pour a few cups of dry sugar on top of the newspaper. The bees can use the dry sugar to meet their carbohydrate needs and don't have to break cluster to feed. The cluster will move as a unit and chew tiny holes in the paper to access the sugar. An added benefit of this method is that sugar is hygroscopic—it absorbs moisture from the air. Any additional moisture left in the hive from late-season feeding will be absorbed by the sugar, which can make it easier for bees to consume the sugar and will help prevent condensation in the hive. If you are mountain camp feeding a top bar hive, simply place the sugar on the floor of the hive.

Other ways to provide emergency feed in the winter include candy boards or sugar cakes made from sugar, pollen, and often additives such as essential oils or vinegars. The internet is full of recipes.

Why Feed Honey Bees?

Let's examine the primary reasons honey bees may need feeding. Understanding the why will help clarify the what, how much, and frequency of feed.

TO MEET NUTRITIONAL DEFICIENCIES

The most common reason to feed bees is to ensure they have the macronutrients necessary for colony activities. As with humans, poor nutrition can result in devastating health effects on the colony. However, humans have decades to try to right any nutritional wrongs; the life span of a bee is measured in days. A few weeks of poor nutrition in the developmental stage can wreak havoc on a colony when those bees are expected to provide the foraging workforce to gather food for the rest of the colony. If the timing is right, and this weakened workforce comes of age during those few precious weeks when bees are expected to make honey, the colony can forfeit its ability to store enough honey for the winter and may starve to death.

Hopefully, the flora in your area will provide plenty of nectar and pollen to sustain your colonies and the need to feed to meet nutritional needs will be a rare occurrence for you. However, even areas that typically present an abundance of resources can experience extreme and unfortunate weather conditions. For example, an extended drought may mean the plants that typically produce the nectar a colony needs to create honey stores may not bloom. Or unusually heavy rains may mean that the nectar gets washed out of the flowers, preventing a colony from storing that nectar as much-needed honey.

Starvation is a real threat to colonies, and one that is totally preventable in managed colonies. Though it is preferable for your colonies to have honey and nectar instead of sugar water to fuel their activities, the fact is that sugar water will prevent starvation if the colony does not have enough resources to keep it well fed.

TO STIMULATE BROOD REARING

Some beekeepers choose to feed a thin sugar syrup to stimulate brood rearing early in the bee season. You have learned how important both pollen and nectar are in the brood-rearing process, and we will discuss how to supply protein and fats if a colony does not have access to adequate pollen. But feeding sugar syrup can cause honey bees to rear more brood than they naturally would. Stimulative brood rearing can result in much stronger colonies earlier in the season, which, if managed properly, may mean a beekeeper can divide their colonies sooner or produce more honey than they would otherwise.

However, proceed carefully if choosing to feed to stimulate brood rearing. Breaks in brood rearing naturally occur during dearths. As resources dwindle, queens slow or stop laying altogether. Breaks in brood rearing can help control for several brood diseases and the reproduction of varroa mites, as you will learn in chapter 10. Additionally, feeding to stimulate brood rearing during seasons when no food is naturally available causes a colony to expand and results in more quickly dwindling food stores. This means stimulative feeding often results in the need to feed to meet nutritional deficiencies. Toying with nature in this way may cause more harm than good, and I therefore reserve stimulative feeding for only during extreme dearths, in cases where the queen has paused egg laying for so long that the entire colony is at risk of collapsing. I recommend that you reserve stimulating brood rearing until you have at least a few years' experience and you better understand how to properly manage the outcomes.

TO PROMOTE COMB BUILDING

Bees can be fed sugar water to encourage them to draw beeswax comb. This is particularly important when trying to establish a colony from a purchased package or nuc of bees. The environment may not provide the nectar necessary to build comb at the time a new colony is introduced. Plus, a package of bees or a nuc colony generally do not have a large number of foragers to collect nectar. If a brand-new colony does not have access to the carbohydrates necessary to build out adequate beeswax comb, the queen will be limited in the number of eggs she can lay, resulting in the colony staying too small to gather the resources needed to overwinter successfully. Providing sugar water to packages of bees is especially important, as upon installation the colony will have no comb for egg laying or food storage. If feeding to promote a new colony to build comb, aim to feed one gallon of sugar syrup per week until the colony has a total equivalent of an estimated ten to fifteen deep frames of comb. (This is a harder estimation when using a top bar. I recommend feeding until your colony has at least ten top bars fully drawn with comb.)

Other reasons beekeepers may feed sugar syrup is to help clear up certain brood diseases, such as European foulbrood. Also, beekeepers breeding queen bees will feed sugar syrup to aid in queen rearing.

What about Pollen?

As you have learned, pollen is crucial for brood rearing. But overall, the need for nectar and honey in a colony far surpasses the need for pollen. Most of your adult worker bees will subsist solely on carbohydrates, and the colony will almost always have far more adult honey bees than brood. Further, only the larval stage of honey bee brood requires feeding. That said, if pollen is lacking in the ecosystem during the brood-rearing months, your queen will adjust her laying in response to the lack of resources, slowing her pace or ceasing egg laying altogether.

A colony's access to pollen will depend upon the flora and climate in an area. In Central Texas, pollen is available almost year-round, so I rarely have to feed pollen substitute to my bees. But remember, even areas normally brimming with resources are subject to the whims of Mother Nature. In February 2021, we experienced an unusually hard and long freeze that lasted more than a week. Spring had already arrived in Texas, and all of our early spring plants had started to bud. The storm killed all the plants that support our bees' early spring pollen needs. Because all our colonies had already started to rear brood in great numbers, pollen substitute was necessary to ensure this brood was well fed for weeks after the storm.

If your region does not supply the pollen necessary for brood rearing during the time when your colonies should be rearing brood and expanding, pollen substitutes may be necessary. Substitutes can be found in beekeeping supply stores in two forms: powdered pollen substitute and pollen patties. If using the patties, cut and insert only a small, three-inch square to the top of the frames or floor of a top bar hive. These patties are a breeding ground for a pest known as small hive beetles, so using a small amount can ensure that the bees consume it before it attracts the beetles. Patties can be made from the powder substitute, or the dry powder can be fed in an open vessel outside the hives. If using the latter method, be sure to place the powder in a covered area where it won't get wet and is out of reach of small mammals, such as racoons, that may want to steal it!

Many beekeepers feed pollen and pollen substitutes as another way to stimulate brood rearing in the early spring. Be careful if you choose this path. Encouraging a colony to rear brood before it would naturally do so can have grave consequences if there is a late freeze and the cluster of bees is not big enough to keep the brood nest warm. Also, the resulting larger colony will consume more honey than it would if allowed to brood up naturally, and the bees will require more feeding if they run out of honey before the nectar flow begins.

RECIPE

Pollen Patties

YIELD: 2

INGREDIENTS
¼ CUP POLLEN SUBSTITUTE
¼ TSP VEGETABLE OIL

DIRECTIONS
Mix pollen substitute and vegetable oil, and add enough sugar water to make the consistency of cookie dough. Place between wax paper sheets and roll out. Cut paper around patties, and store in freezer until ready to use.

Also keep in mind that if the bees have access to true natural pollen, they will likely ignore any pollen substitutes. In which case, stop feeding it and save yourself the money!

How to Feed Honey Bees

Let's discuss a few of the more common feeders and the pros and cons of each.

FRAME FEEDER

A frame feeder is used in a Langstroth hive and takes the place of one or two frames in the hive body. Frame feeders are available in both deep and medium sizes to be used in two different hive body sizes. I prefer frame feeders because they tend to hold a decent amount of syrup (usually one gallon) and the feeding takes place in the hive, so it doesn't encourage robbing by other bees. The downside to this feeder is it requires opening the hive to check the feed level and add more syrup. If choosing a frame feeder, I strongly recommend one that has what are known as caps and ladders. These are accessories to the frame feeder that help prevent drowning bees. Without them, frame feeders can be a death trap when filled with syrup.

BOARDMAN FEEDER

Also called an entrance feeder, this feeder is intended for use at the entrance of a Langstroth hive or on the floor of a top bar hive. It has a tray, comes with a lid with small holes punched into it, and is intended to go on a mason jar. This feeder allows bees to suck syrup through the holes. The greatest benefit of the boardman feeder is that the jar is visible from the outside of a Langstroth hive, so you can check the levels and change out the feed without getting into the hive.

However, many beekeepers avoid using boardman feeders because they can incite robbing. If robbing is common in your area, you may want to avoid this feeder type. For example, here in Central Texas robbing bees isn't generally a concern in the spring months, but I discourage beekeepers from using these feeders in the fall, when robbing is quite common. To mitigate this risk, you can place the feeder on top of the inner cover of your Langstroth hive with an empty hive body over the feeder, then place your inner cover and telescoping cover on top of the extra hive body. This prevents access to the feeder from robbing bees and wasps.

My second complaint about this feeder is that the capacity is dependent on the jar size, and the most readily available jars are quart size. This means if aiming to feed a colony one gallon of syrup each week, the jar will need refilling four times over the week. Of course, glass jar feeders can also topple and break, especially if the neighborhood raccoons get involved.

TOP OR TRAY FEEDER

Several types of feeders are designed to sit on top of a Langstroth hive. These are my least favorite feeders for several reasons: First, their capacity is usually very large, sometimes four gallons or more. I recommend against feeding a colony more than a gallon per week, and this larger capacity can encourage a beekeeper to ignore this recommendation. Second, they can be challenging to move when filled with syrup, and spills are almost inevitable. If an inspection is necessary when the feeder is full of syrup, be very careful when removing the feeder. Plus, they are big and bulky, and they can be cumbersome to store.

OPEN FEEDING

Open feeding describes feeding syrup away from the hives so that the bees must leave the hive to gather syrup. I strongly recommend against open feeding for several reasons. First, open feeding results in feeding every other bee in the neighborhood! Bees within several miles will come to steal the syrup. Second, the nature of open feeding means you can't guarantee the colonies that most need the feed will get the syrup. The strongest hives will take most of the feed before the weaker colonies can access it. Third, it's impossible to gauge how much feed a colony is consuming with open feeding, which is critical data I track when feeding bees. And finally, open feeding will cause a robbing frenzy among colonies, which can kill a large number of bees and assists with the spread of pests and disease. If you do choose to open feed, make sure to site your feeding station at least thirty yards from your own hives to prevent the robbing bees from turning on your colonies next!

A Few Best Practices

Here are a few best practices to consider when choosing to feed honey bees.

1. Never feed bees more than they can consume in one week. Sugar syrup can ferment, and once it ferments the bees ignore the syrup. It will then become a breeding ground for small hive beetles. If an inspection finds syrup remaining after one week of providing, dump the sugar syrup (cleaning the feeder if necessary) and adjust the amount you provide for the next feeding.

2. More mouths = more feed. A strong colony will easily consume one gallon of sugar syrup each week, but a weaker colony will struggle to consume that much. Be sure to adjust your feed accordingly. Also, stronger colonies mean more mouths to feed, so whether sharing honey or feeding, dedicate more of your resources to these hives.

FEEDER TYPES AND WHERE THEY GO

1 FRAME FEEDER

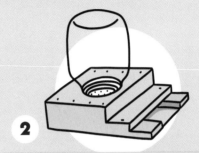

2 BOARDMAN/ENTRANCE FEEDER

3 TOP/TRAY FEEDER

4 OPEN FEEDING

(not recommended)

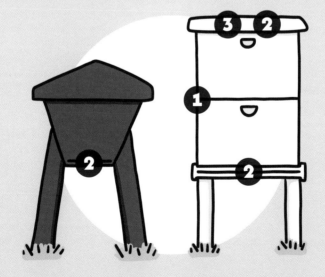

3. Do not feed your bees more than a gallon of sugar syrup per week. Bee-keepers often fall into the trap of assuming that if bees are taking the syrup, they need it, and so they feed as much as the bees will consume. However, because bees only have access to food for a short period of the year, they have evolved to become very effective hoarders. They will continue to store syrup and nectar that is available as long as they have the bees to do it. This is a surefire way to cause your colony to become "honey bound." A *honey-bound colony* is one in which the bees start filling the brood nest with nectar, honey, or, in this case, sugar syrup. This is also called backfilling the brood nest. Backfilling the brood nest also can cause a colony to swarm, as the queen no longer has space to lay eggs.

4. Feeding is not a silver bullet. You cannot solve all the problems of a sick or weak colony by feeding. The fact is that feeding a queenless colony, one in-fested with varroa mites, or a colony that has grown weak will not fix these problems. Review the reasons to consider feeding, and feed only if those are your concerns. Overfeeding a colony that has grown weak, believing it will make the colony grow stronger, will simply divert the bees' attention away from brood rearing, which is what the weak colony actually needs.

5. Lean in to your local bee community. I can't tell you exactly how much honey your colony needs to survive the winter because this is a very region-specific consideration. Here in Central Texas, thirty to thirty-five pounds of honey is sufficient, but beekeepers in the northern part of the United States usually need at least double that amount. This is why it is important to find a local community, whether through a mentor, a local bee club, or a professional beekeeper who does consults or classes in your area, to help determine your target honey goal. Furthermore, a local community will share when you can generally expect your nectar flows to start.

6. Let your colony and your philosophy be your guide. Understand the status of your colony before making a decision of whether to feed. Inspect your hive to understand what is happening in your colony, then assess if supple-mental feed is necessary. Advice that you "should" be feeding a colony from someone who isn't aware of the circumstances in your hive isn't worth a hill of beans.

Can You Overfeed a Colony?

If a colony does not have access to proper nutrition, the colony can and will fail. On the other hand, too many beekeepers with a poor understanding of the needs of bees often treat them as pets that require daily feeding, stuffing the colony full of pollen substitutes and sugar water. Though well intentioned, these beekeepers' colonies can experience grave consequences as well. A colony can become honey-bound and swarm, and overfeeding a weak colony diverts resources away from brood rearing. Excess syrup or pollen patties left on a colony that the bees cannot consume in a week can become a breeding ground for small hive beetles. Additionally, although table sugar is the closest substitute we have for nectar and honey, the pH level of sugar syrup and honey are different and pathogens grow more readily at the pH level of sugar syrup. And feeding too early in the season to stimulate brood rearing can cause brood to perish in the case of a hard freeze, when the cluster can't adequately keep the now large brood nest warm.

CHAPTER 8

All Things Swarming

Colony Reproduction

Swarming is the colony's way of reproducing. Unfortunately, "swarming" is one of the most misused terms related to honey bees and beekeeping. Swarming does not refer to a cloud of attacking bees, and swarming is not what a colony does when it's unhappy with its abode and leaves looking for greener pastures: this phenomenon is actually called *absconsion*! A colony that is strong and has enough resources to reproduce does this by casting a swarm, hopefully creating two colonies out of one.

Remember that we need to think of the colony as a whole as an organism. And like all living beings, this organism needs to reproduce. The mere mention of a swarm can put simultaneous fear and excitement in the heart of a beekeeper. Most beekeepers would prefer that their own bees not swarm, yet catching another colony's swarm can mean "free bees!" In this chapter, you will learn the ins and outs of how a colony swarms and how to prevent swarming in your own apiary, and how to bait and catch your own swarms.

What Happens When Honey Bees Swarm?

Several days before a colony swarms, the queen bee stops laying eggs and is put on a diet by the worker bees. This diet decreases her body weight and allows her to fly with the swarm, and the worker bees stuff themselves full of honey to take on their journey. On the fateful day, roughly half of the young worker bees will leave the colony en masse with the queen. The swarm is looking for a new home to start a new colony.

This cloud of bees is an amazing sight to see, and it looks quite chaotic and disorganized. However, it's actually the opposite: though the swarm has not yet selected a new home, they will very quickly organize together, a sort of ball of bees, on some object. It's from this temporary site that the scout bees will begin the hunt for new real estate. The swarm will rest in a tight ball, with the queen protected at the center. This swarm may rest on a tree, a fence, a car, or even a mailbox!

Once settled, scout bees will begin flying from the swarm, looking for a suitable home. Just as bees use the waggle dance to communicate the location of food, they also use it to communicate that of potential new homes. A scout bee will move her body in a figure-eight shape across the comb, and when she reaches the middle of the figure eight, she will shake her body rapidly. The angle of the dance communicates the angle of the sun to the potential new home. The duration of the waggle communicates distance: the longer the waggle lasts, the farther the new home is from the swarm site. Next, each of the scout bees will visit all of the potential homes, return to the swarm, and "vote" for their favorite by performing the waggle dance corresponding to their favorite new home site. When all of the scouts are performing the same dance and therefore are in agreement, the entire swarm will move again, this time to the new home. This house hunting can take as little as a few hours or as long as a few days. The bees that have engorged themselves with honey will immediately begin to use the carbohydrates to build the beeswax comb so the queen can again start to lay eggs and establish the new colony.

The bees that are left behind in the original colony began their preparations for a new queen before the swarm ever left the hive. A few weeks before, they started rearing several queen bees, which will reach maturity and emerge after the swarm is gone. However, the first queen that emerges will usually sting and kill all the other developing queen bees still in their cells! This lone victorious queen bee must now go on a mating flight before she can start laying eggs in the hive.

Lessons from a Beekeeper

Every beekeeper who's been around a while has one good swarm story. Catching swarms is one of my favorite activities. We usually get our first swarm call by early March, and it aligns with the burst of spring onto the scene. It embodies the excitement of the season to come, and those first few calls remind us that our big season is almost here. The most challenging part of swarm catching, and what makes it so adventurous, is that swarms can be found in some interesting places. Often, the hardest part of swarm catching is finding a way to get close enough to get your hands on the swarm! I'm going to share two of my favorite swarm stories, and if you stick with beekeeping, I have no doubt that you'll soon have your own to share!

THE ROOMMATE RIDE-ALONG

I received a call from a student who excitedly reported a swarm in her back-yard tree. It was the first call of the season, and I grabbed my tools. I had a roommate at the time, and just that morning he had been complaining that he felt trapped: his days felt monotonous and adventure was lacking. I had a proposal—was he truly ready for an adventure? Because if so, I sure had one for him. He didn't skip a beat before jumping in the truck. When we arrived, we found the swarm very high in the tree, at least eighteen feet. We didn't have an extension ladder tall enough, and our A-frame ladder could get me up only about eleven feet off the ground safely. We strategized and hypothesized and brainstormed. The swarm was too far away from the trunk to shimmy up, and even if we did, the branch wouldn't hold our weight. Finally, an idea! We would throw a rope over the branch and tug down gently to get the branch a few feet closer. We did a sort of mock-up of the plan. I stood on the top step of the ladder (you know, right where it says "do not stand") and my roommate held my feet to keep me steady. I asked for a bucket to hold stretched over my head, and as my student's husband gently tugged the branch down, we realized we could get within a few feet of the branch! Excited that we had worked out a plan I yelled down, "OK! Let go!!" As I watched the branch bounce up, my brain started piecing together that this was meant to be a practice run. Only one person in the yard was wearing gear, and it wasn't me, nor my roommate holding my feet down below. The branch sprung up, and the swarm came falling down into the bucket, with bees crashing all over my face and my roommate's head. I looked down, terrified about what his reaction might be. He had about a dozen bees all over his face and head, and he exclaimed, "Are they stinging!?" I replied, "Does it hurt? No? Then they aren't stinging!"

The irresponsibility of my actions isn't lost on me, but thankfully this swarm was like almost every other I have encountered—very docile. I climbed off the ladder and shook the swarm into the student's hive. That swarm continued to be very productive for her for several years. And the best part is that everyone that day got an unexpected adventure! The lesson: always, and I mean always, ensure that everyone is wearing protective gear on a swarm call . . . even on the practice run!

THE FIRST DATE

Later that same year, I was on a first date, and during dinner my phone dinged. A quick glance told me I had a swarm call. Now normally, I wouldn't interrupt a social function for a silly swarm, but the swarm was in one of my own apiaries. I feared it was my own hive swarming, and that is definitely worth cutting a dinner short. I profusely apologized to my date but explained that I had a bee emergency and had to go. I showed up, and sure enough, there was a swarm on the bottom of a lawn chair. I didn't have gear with me, and despite the stupidity of handling bees at night and without gear, I proceeded. One of the residents whipped out a phone to catch me on film doing a swarm capture in a sleeveless mini dress and three-inch heels. I picked up the chair, shook the swarm into a hive, and was done in less than five minutes. The video is still a favorite, and I share it each year during swarming season.

Are you wondering whatever happened with the fella who got his dinner date cut short? It didn't work out. Though he is truly one of the nicest guys I've ever met, somehow I couldn't shake the fact that he never seemed interested in coming along for the adventure. The lesson: never, and I mean never, catch a swarm at night in a mini dress and heels. And if the person you're interested in is a (bee) keeper, tag along for the swarm catch. . . . No doubt you'll win a few extra points.

It's easy to see why beekeepers usually want to prevent their hives from swarming! Not only does the beekeeper lose half of a colony, but the remaining colony is left weakened, with less honey, and queenless. Plus, much can go wrong in the requeening and mating process, resulting in a queenless colony unable to rear young.

It is not unusual for a hive to swarm multiple times, and in fact, the colony has prepared itself for that scenario. As the workers begin to rear queen bees to replace the original queen mother, they will rear multiple queens over multiple days, staggering the development of the queen bees. If the first queen emerges and the colony is still quite strong, the workers may decide to cast off a secondary swarm. The colony will swarm again with the newly emerged queen, and the remaining developing queen bees will be left behind intact. Swarming can happen multiple times, and I have seen colonies almost swarm themselves to death, leaving barely enough resources behind to ensure the parent colony can still survive. The first swarm is referred to as the primary swarm, next comes the secondary swarm, then the tertiary swarm, and so on.

WHAT CAUSES A COLONY TO SWARM?

Though scientists are still working to understand the conditions and communications that trigger swarming, we can learn from the current research and anecdotal observations of beekeepers who have worked to prevent swarming since colonies were first managed, thousands of years ago. In a colony, the first step in the swarming process is the rearing of new queens. This queen rearing is necessary to ensure replacement of the original queen, which will leave with the swarm. Therefore, to better understand what causes a colony to swarm, let's examine what conditions trigger this first step to swarming:

- **Colony size:** Rapid population growth is one signal to the colony to begin rearing new queens. Once a colony reaches a certain size and population of bees, it is more likely to begin swarm preparations.
- **Congestion in the hive:** Congestion in a hive signals that swarming may be possible. The queen mandibular pheromone inhibits the initiation of queen rearing by worker bees. These pheromones are spread through the colony by the worker bees touching one another after they have touched the queen. As a colony grows in size and becomes more congested and packed with bees, the spread of the QMP is inhibited, and queen rearing is triggered.
- **Constraints on egg laying:** As a colony grows rapidly during swarming season, it may reach a state where there are enough workers to support additional brood but the queen cannot produce eggs quickly enough or there are no more unused cells in the brood comb. Once this state is reached,

LET'S

THE BEES THAT SWARM:

1 The worker bees stuff themselves with honey . . . but the queen is put on a diet.

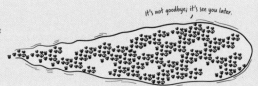

2 1/3 to 1/2 of the bees leave.

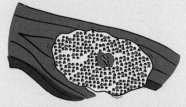

3 The swarm will rest in a tight ball with the queen at the center while scouts go househunting.

4 The scout bees fly and look for a new home.

5 They return and perform the waggle dance until all scouts agree on a new home.

6 On move-in day, the worker bees use the honey stores to immediately begin making new beeswax comb.

SWARM!

' We need more eggs!

This hive has too many young 'uns.

THE BEES THAT STAY:

1 The worker bees start rearing several queens.

2 If the colony is strong enough, they may choose to swarm again.

3 If not, the first queen that emerges will kill the other developing queens.

4 Then, she'll go on a mating flight in search of drones.

Come & get it, boys!

(Much can go wrong here. See "Ways to Kill a Queen" in chapter 6.)

5 Should all go well, she'll rejoin the hive and start laying eggs.

the only way to satisfy the evolutionary goal of continuing reproduction is to produce more queen bees. A strong nectar flow or overfeeding of sugar syrup by the beekeeper also can contribute to this limitation of space for egg laying as bees begin to store the food in cells that would normally be used for brood rearing. As you learned in chapter 7, this is what is known as becoming honey bound.

- **Worker bee age distribution:** The proportion of young to older bees can initiate swarming. As a colony comes out of winter at the start of the season, the age distribution of a colony is skewed toward older worker bees. This is because the queen has slowed or halted her laying of new brood altogether during the colder months. As the queen begins to increase her laying, the population of younger bees increases rapidly. Once this proportion of younger to older bees reaches a certain threshold, queen rearing is triggered.

PREVENTING SWARMING

Just as talk of catching a swarm (free bees!!) puts a hopeful smile on the face of any beekeeper, the thought of their own colony swarming can bring about looks of dread and disappointment. Know that in trying to prevent swarming, we are working against the will of nature, where reproduction is always equated to success. Because swarming is the superorganism's way of reproducing, a colony will try to swarm if the conditions are right. Understanding the conditions that will trigger the rearing of queen cells allows us to better prevent these conditions in our apiary. Here is a list of suggested steps to take if you sense that any of the previously described conditions are true.

Opening up the Brood Nest

First and foremost, make sure the hive has adequate space to house the bees and the queen has plenty of space to lay eggs. How to do this can look a bit different in different hive types. In a Langstroth hive, this means adding drawn frames of comb to the hive, and it may mean adding hive bodies. In a top bar, this may mean providing more top bars of drawn comb and moving your follower board to give your bees access to more of the hive. However, not all space is created equal. Remember that honey bees cannot build beeswax comb, which is necessary for the queen to lay more eggs, unless they have access to ample carbohydrates. This means that if your bees are not bringing in nectar and building new comb and you need to give your queen more space for eggs, you must give the colony drawn comb from another hive or from storage. Simply giving a hive a frame or top bar without drawn comb does not give the queen new space to lay: drawn beeswax comb is a must here.

The act of adding more space for the queen to lay is called opening up the brood nest. If you are a newer beekeeper, you may not have a lot of drawn comb to spare, in which case keep reading for other ways to reduce congestion and open up the brood nest. You can also experiment with feeding sugar syrup to encourage drawing comb to provide the queen more space to lay.

Sharing Brood with Other Colonies

Colony size and congestion in the brood nest is a contributing factor in swarming, and this congestion may be relieved by sharing frames of brood with other colonies. This is also a great way to boost weaker colonies, as the receiving colony will accept and rear the brood as their own. If possible, look for a frame of capped worker brood, or pupae, to share from the stronger to the weaker colony. These bees are the closest to emerging, and this will do more to ease congestion in the stronger colony and more quickly add new worker bee resources to the weaker colony. The brood will need to be traded with a frame of their own. You will need to swap a frame from the receiving colony: try to find a frame of drawn worker comb that is empty, which will give the queen in the stronger colony more space to lay eggs. (Remember, worker bee cells are smaller than drone cells, so look for drawn comb full of the smaller worker bee hexagonal cells.) Place this frame of drawn comb in the brood nest. Make sure to not share a queen bee in the process! Check the frames for queens, and with a forceful downward motion, shake off most of the bees over the original colony before the swap.

Making Increases or Splits

A beekeeper can manipulate colonies on the verge of swarming and create new colonies for their bee yard. This is called making increases, divides, or splits, and is a great way to grow an apiary! By taking resources of brood, honey, pollen, and bees from one or more hives, you can create additional colonies and keep those "swarms" in your own bee yard. Making increases eases brood congestion and helps balance the ratio of young to older bees in a colony.

Other Ways to Ease Congestion

Removing entrance reducers and adding entrances also can help ease congestion. Top bar hives often already have multiple entrances, and you may just need to open those extra entrances. In a Langstroth hive, you can add a second entrance by drilling a hole in an upper hive body. An easier, and less permanent, way of adding a second entrance is to slightly offset the top hive body from the hive body below it to provide a one-inch opening at the top of the hive.

All this said, swarming is very likely to happen in your apiary at some point.

When a colony swarms, just remind yourself that in the eyes of your bees, they have just accomplished a great feat! Give yourself a congratulatory pat on the back for raising successful bees that completed their evolutionary goal!

Catching Swarms

Catching swarms is a great way to increase your apiary size with "free bees." Plus, helping to house swarms can contribute to the greater good of honey bees. Not all swarms will survive: a swarm has the daunting task of building a new home and storing enough honey before winter comes.

If you get a lead on a swarm, ask whether someone can safely take and share a photo or video of the swarm. Most of the uninitiated don't actually know what a swarm of bees looks like. I have received swarm calls for wasp nests, colonies already nesting in trees, and even just a few dozen bees feeding off sugary drinks in a trashcan. You don't want to pack up and head out only to be disappointed when you arrive. Also, ask the tipster to keep an eye on the swarm and let you know if they leave. A swarm may hang in place as little as an hour or as long as a few days, so if they do leave, a courtesy heads-up is nice so you don't waste your time.

A swarm kit should include, at a minimum, your smoker, protective gear, and a vessel to hold the swarm. Make sure the vessel is ventilated so the bees don't suffocate, but is fully enclosed so that bees cannot escape the box. This could be a bucket with holes punched in the lid, a nuc box with an entrance that can be shut, or, in a pinch, even a cardboard box with small holes punched in the sides. You also may need tools to help access the swarm, such as a ladder. Other items that may be helpful include a queen clip (a small contraption that allows you to capture and house your queen bee), a sprayer filled with sugar water, and a white sheet.

Once you arrive and find the swarm, light your smoker and don your protective gear. Avoid using the smoker if possible, but have it handy if the swarm gets uncharacteristically aggressive. Using a bee smoker interferes with the worker bees' ability to find their queen, and your primary goal in catching this swarm is to capture the queen. Once you do, the workers will perform the rest of the work for you! The worker bees, following the QMP, will walk right into the vessel. Though rare, a swarm can be aggressive, so make sure the smoker is lit if needed.

The secret to swarm catches is finding the queen bee! The queen will be located in the middle of the swarm, so knocking or brushing as much of the swarm into the vessel as possible is the best plan of action. Lightly spraying the swarm with sugar water first allows the bees to drop into the box more easily and prevents them from flying. Once you get as many bees in the vessel as possible, stand

Why Not Let Your Colony Swarm?

Some beekeepers do not expend efforts to prevent swarming. These folks often also tend to have a very hands-off approach to beekeeping, and rather than enjoying the experience of the actual beekeeping, they enjoy the presence of the bees from afar. Beekeepers often refer to this category of beekeepers as "bee-havers," and the tone is usually intended to be a derogatory one. I strive for inclusivity and believe there is plenty of space for all types of beekeepers. However, if you are interested in the process of beekeeping and hope to harvest honey someday, I do encourage you to consider learning more about how to prevent swarming. Let's discuss the reasons.

1. A colony that swarms may be unsuccessful in requeening itself. If a beekeeper does not intervene to provide a breeder queen bee or worker bee eggs (and therefore an opportunity to rear a new queen), this colony will diminish quickly, eventually failing.
2. A swarming colony likely will not provide a honey harvest that season. A colony that swarms loses much of its honey to the swarm and must rebuild its population after the loss. Of course, there's more to beekeeping than honey, but knowing that a colony that swarms may not provide any harvest can be a good motivator to prevent swarming.
3. Splitting a colony, instead of allowing it to swarm, is a great way to increase your number of colonies without buying more bees.
4. A swarming colony means losing a productive queen bee!
5. Finally, if you live in a more urban area where houses are close together, preventing swarming is good for neighborly relations. While a swarm is a beautiful sight to see, it can be disturbing to those who don't understand the mechanics of the swarm. Also, this swarm may end up building a nest where they aren't wanted: in the eaves of your neighbor's house, for example.

Those with a more hands-off approach have their own reasoning for allowing their colonies to swarm, and everyone has to develop and follow their own philosophy. However, I recommend that beekeepers protect their investment and keep it from flying away, with a proactive approach to swarm prevention.

back and observe their actions for several minutes. The bees will congregate where the queen is located. This is why a white sheet is helpful: laying the sheet below the swarm will allow you to easily find the queen if she ends up on the ground. The white background makes it easy to spot any congregating bees. Once you have captured the queen and most of the bees, close up the vessel and bring the bees to your apiary to house them in their new home: one of your hives!

Know that swarms can be very flighty and finicky, and they can abscond after install. You must take a few extra precautions to entice the colony to stay in the hive. Here are some tips:

1. Share a frame of open brood from another colony with the swarm. The swarm won't want to abandon the brood.

2. Use a queen excluder to keep the bees inside of the hive. A queen excluder is a metal or plastic tool that prevents the queen from entering certain areas of the hive but allows worker bees to pass through. You can use a queen excluder to keep the queen, and therefore the swarm, from leaving the hive. In a Langstroth hive, queen excluders are flat pieces that can be inserted between hive bodies. Place the queen excluder between your bottom board and the first hive body. Top bar hives have circular disks that act as queen excluders. Place the queen excluder disk over one entrance and block off all other entrances. Don't leave the queen excluder on for more than a few weeks, as drones also cannot pass through an excluder. Unable to exit, they will die inside and the worker bees won't be able to remove them from the hive.

3. Finally, avoid disturbing the swarm for at least a week. Doing so can make the swarm feel threatened or unsafe, and the bees may abscond.

Baiting Swarms

In addition to catching swarms, beekeepers also can take a proactive approach and set traps to try to bait a swarm! I recommend baiting swarms in your apiaries, because if efforts to prevent swarming in your colony fail, you may catch your own swarm. I strongly recommend using a trap that can house the frames or top bars you use in your own hives. That's because if a swarm does move in, you can easily transfer the frames or top bars over to the hive. We use six frame wooden nuc boxes as swarm traps, but you can find plans online to build your own swarm trap. A single hive body with solid boards added to make a top and bottom, and a drilled hole for an entrance, also works well. To make the trap feel welcoming and enticing, use a frame of drawn comb in the trap, and use a lure to make the trap "smell" familiar. You can buy swarm lure from a beekeeping store or making your own. I've included an easy-to-make recipe we use in our own swarm traps. Simply apply a teaspoon or two inside of the swarm trap and another at the entrance, and reapply every two to three weeks. Using hive bodies and nuc boxes that previously housed bees has an additional benefit, since the odors from the previous colony are enticing to a swarm looking for a place to live.

RECIPE

Swarm Lure

YIELD: ~¼ CUP

INGREDIENTS

2 TBSP BEESWAX
¼ CUP OLIVE OIL

40 DROPS
LEMONGRASS OIL

DIRECTIONS

Heat beeswax and olive oil together. Remove from heat. Add essential oil. Stir and store in jar or tin.

CHAPTER 9

Growing an Apiary

Making Increases

Colonies reproduce through swarming, and during this process a strong colony divides itself into two colonies. In the previous chapter, you learned about the pitfalls of your colony swarming along with ways to prevent it. The upside to bees' inclination to swarm is that a beekeeper can take advantage of this natural process and, if done correctly, can prevent the colony from swarming and grow their apiary at the same time! We call this *making increases*, though you'll also hear beekeepers use other terms to describe this process, including making nucs, splits, and divides. We will get into the semantics shortly.

Even experienced beekeepers tend to shy away from making increases, and many find themselves intimidated by the idea. However, if you understand the needs of a colony, it is quite simple to understand the mechanics and steps involved in making increases. There is no shortage of types of increases and methods to use, but at the end of the day, the needs of the colony are always the same. It doesn't much matter which method you choose as long as both colonies are left with the resources to meet their needs.

The reasons for making increases are plenty. Depending on the time of year and method used, beekeepers can use splits to reach a variety of goals. Beekeepers may make splits to:

- Increase their apiary size and number of colonies
- Increase honey production
- Decrease honey production (splitting a hive in a way that decreases its workforce means a colony will make less honey if a nectar flow will occur soon)
- Prevent swarming
- Keep colonies smaller and easier to manage
- Raise queens
- Produce nucs to sell to other beekeepers

The Needs of a Colony

Increases are not possible in every apiary in every hive or in every year. Some beekeepers misunderstand and believe if they start with two colonies in year one, the colonies will continually need to be split each year, leaving them with four colonies in year two, eight colonies in year three, and so on. If it were that easy, we beekeepers would be making a lot more money. (I'd be on a beach somewhere instead of writing this book . . .)

If you are planning to make an increase in your apiary from one or more colonies, you must provide the original and all new colonies with four resources:

1. **A queen bee:** Every colony must be given a queen bee or the ability to rear a new queen. This can mean buying a queen from a breeder for the new colonies or giving the colonies a frame of eggs to rear their own.
2. **Adequate nutrition:** Every colony must be left with an adequate supply of food. This can mean sharing honey stores with a new colony or feeding it sugar water if honey or nectar is not available.
3. **Adequate brood:** All colonies must be left with brood to ensure a continued supply of bees to keep the colony functions going. Because worker bees die every six to eight weeks, the colony needs brood to ensure future generations of bees. This is particularly important if you plan to let one of the colonies requeen itself.
4. **Adequate nurse bees:** Finally, make sure there are enough nurse bees in each colony to completely cover each of the frames of brood. These nurse bees are necessary to keep the brood warm and continually feed the larvae. Without enough nurse bees to cover the brood cells, the developing bees will die. One nurse bee is required for every two brood cells. This is where most beekeepers go wrong in making their first splits, and we will cover this in greater detail.

4 NEEDS OF A COLONY

✓ a queen bee

✓ adequate brood

✓ proper nutrition

✓ nurse bees to care for the brood

Use only your strongest hives to help make splits: do not try to make increases from weaker colonies. Of course, you will learn that you need not take all the resources for a split from a single hive—you can use multiple colonies to provide the resources to make increases.

When to Make an Increase

Certain conditions must always be met before a colony can be split, and these conditions are not true every year and may only be true in certain seasons of a given year. Do not make a split before queen bees are available for purchase or, if you plan to let a split rear its own queen, before sexually mature drones are present in your area.

Because queen bees must mate with drones before they can lay eggs, and drones are not an available resource year-round in most areas, queen bees are not always readily available for purchase. Queens are available for longer periods in warmer climates, such as the southern part of the United States.

Usually, your colony will have the ability to rear a queen that can mate with area drones before you can buy a queen locally. This is because queen breeders usually use the first queens their apiaries produce to replace queens in their own hives and to place into packages of bees and nucs. If you plan to make a split very early in the spring season, you may be forced to allow your colony to requeen itself rather than buy a queen from a breeder.

However, be certain your area will have sexually mature drones readily available by the time your queen will be ready to leave the colony to mate. Drones are sexually mature approximately sixteen days after emerging.[1] To determine whether you have sexually mature drones in your area, a little deductive reasoning is necessary. It takes twenty-four days from the time a drone egg is laid before the drone will emerge, and another sixteen days before that drone is sexually mature. Therefore, I can assume a colony is about forty days away from having sexually mature drones once I spot drone eggs. Of course, the drones in a hive aren't necessarily the ones that will mate with the virgin queen from any split created, but I use what's happening in my existing hives as a benchmark to predict when it is happening in other colonies in any given area. How early sexually mature drones will be available in any given area will depend on that area's climate. In Texas, sexually mature drones are usually available around early March, though this can vary even from one year to the next in any given area. This is because queens do not start rearing drones until an abundance of pollen is available.

Also, be sure that any new colony you create has plenty of time to prepare for winter. If your apiary has lots of resources to share with the new colony, you can safely make splits later in the year. If the new colony will be shorter on resources—brood, bees, or honey—you will want to make these splits earlier in the season to give the colony time to grow and store enough food to last the winter. Generally, be cautious if you are trying to make splits after your primary nectar flow has ended.

The Mechanics of Making an Increase

An increase can be as simple as opening a hive and taking frames or top bars of brood, bees, and honey to place in a new hive. As long as you're meeting the four needs discussed previously, pick and choose what resources you have to spare from existing colonies to provide for the new colonies. A few things to keep in mind:

- Resources can be pulled from multiple hives. This means that you can pull frames of brood from two or more hives, and pollen and honey from yet another hive.
- Each colony should be left with or given a new queen, or given a frame of eggs. You can leave the original colony with a queen or take the queen into a new split, leaving the original colony with a frame of eggs to rear a new queen.
- When taking bees from multiple hives into a new colony, do not combine bees from one colony into another that has an existing queen. If you plan to combine bees from different colonies, you must either give them a frame of eggs to rear a new queen or give them a caged queen and allow them time to adjust to her before releasing the queen. If you introduce a large number of worker bees into a colony with an existing queen, they may kill the queen! You can, however, share frames of brood from different hives without any concern about acceptance.
- Bees from two hives combined together will fight some, but bees from three or more colonies tend to have less infighting. In all the confusion caused by the mix of pheromones, they tend to work it out a bit sooner!
- Unless you move the new split several miles away from the original colonies, any foragers who leave the hive after the split was made will return to the original colony's location. This does not mean you must move the new colonies off site. It simply means that you must be certain to take more than enough nurse bees into the new colonies to account for the bees returning to the original colony.
- In the same vein, unless you move the original colony off site, the new colony will not have any foraging bees to collect food for the hive. Make sure

you provide adequate nutrition for this new colony until the nurse bees can age into foragers and can provide for themselves.

- Bees don't wear name tags, so it can be hard to understand how to make sure you get enough nurse bees in a split. Nurse bees hang out on the brood frames, so if you are taking brood frames completely covered in bees, you are likely getting mostly nurse bees in your new colony.

- Be sure all the brood frames in a new split are completely covered with bees. If they are not, go back to one of the original colonies and shake bees off of a few additional frames of brood. This acts as a sort of insurance policy to make sure you have plenty of nurse bees for your split. Be mindful of where the queen bee is! Don't take her into the split unless that was your plan from the start and you're sure you're not introducing bees from another colony to the mix.

- For your new colonies, aim for a balanced colony with a slightly higher proportion of capped brood. Capped brood requires less care and will help ensure that your new colony has new nurse bees more quickly than if you took all young brood.

Whether you plan to let a colony requeen itself or purchase a breeder queen, I recommend that you review chapter 6 before you make your first split. There, you can refresh your understanding of how a colony produces a new queen bee and how to introduce a new queen to a colony.

Types of Increases

There is no shortage of methods for making increases, and most experienced beekeepers have their favorite way of making splits. The following are just a few types of increases. As long as you make sure each colony meets the four needs we have described, the method by which you achieve the split doesn't matter much.

SWARM CONTROL SPLIT

This is a split for a very specific scenario, in which a colony has begun to prepare to swarm but has not yet swarmed. This means the original queen is still in the colony, but the colony has begun to rear new queen cells and is a few days away from swarming. Once swarm preparations are underway, a swarm control split can keep both colonies in your apiary. In this split, you are essentially performing the swarming for the colony. The split receives the original queen and enough resources to satisfy all the needs of the new colony: brood, bees, and food. *Do not take any of the queen cells.* In the original colony, leave enough resources to satisfy the colony, along with at least one queen cell. *Do not leave the original queen.* If done properly, both colonies "believe" they have swarmed, and this will ensure your bees stay where you want them—in your apiary! Either let the original colony proceed with requeening itself or provide a breeder queen. If you choose the latter, cut down all queen cells before installing a new queen. Keep in mind that if the colony has already swarmed, you will be unable to make this split, as the original queen has already left with the swarm. You must wait to see whether the original colony requeens itself.

WALKAWAY SPLIT

A walkaway split is one in which the colony is left to rear its own queen. The benefit of a walkaway split is you don't have to find the queen in the colony, unlike in a swarm control split. Simply divide the resources of the colony, making sure both the original and the new colony have at least one frame of eggs, and come back in thirty days. If all goes as planned, by that time the colony that was left queenless should have reared a queen that has completed her mating flights and is now laying eggs in the hive. If choosing this split, move the original colony to a new area in the apiary and leave the new colony in the location of the original hive. By moving the entire colony, all the nurse bees will stay with the original colony. Leaving the new colony in the original location will ensure that all the foraging bees will return to the new (and typically weaker) colony.

OVERNIGHT SPLIT

An overnight split is a great option for a beginner who is nervous about making sure they take enough nurse bees into a split. The downside to an overnight split is that it requires two visits to the bee yard. Overnight splits require the use of a Langstroth queen excluder, so they can only be done in Langstroth hives. Open the hive you want to split and pull three to five frames of brood from the brood nest, being sure to replace with other frames. (How much brood you will take is a matter of preference and how much brood the original colony can spare. You also can take brood from other colonies for the split.) Shake or brush all the bees on the brood frames back into the hive. Place a queen excluder on top of the uppermost hive body, then place the brood frames in a hive body on top of the queen excluder and close the hive's inner and top cover. Overnight, nurse bees will come up through the excluder to take care of the brood, but the queen bee will be unable to move to the top hive body. On day two, you can move the top hive body with the brood and all the nurse bees into your new hive in its new location. Please keep in mind that you must satisfy all the other needs of the new colony: be sure to take plenty of food and a frame of eggs so it can rear a new queen, or plan to provide it with a caged queen bee.

NUCLEUS COLONY

Just as you would buy a nuc from a breeder, you can make your own! Take three or four frames of brood, one or two frames of honey and pollen, and plenty of nurse bees, and provide it with a new queen or a frame of eggs. Remember that the resources can come from more than one hive. A nucleus colony by definition has fewer resources than a full-strength colony, so ensure that its needs are met by feeding it or continuing to share brood with the colony so it can grow over time.

Combining Colonies

Combining colonies is a great tool if one colony goes queenless and is unsuccessful in requeening itself. Beekeepers also often combine colonies in preparation for winter: it's wise to combine weak colonies in the fall to help ensure that they are strong enough to survive the winter, then make splits as allowable in the spring.

In combining colonies, the pheromones from the two colonies must be allowed to mix to prevent infighting and to ensure that the bees do not try to kill the queen in the other colony. When preparing to combine two colonies, make sure only one of the colonies has a queen. If both colonies are queenright, choose the queen that is less desirable—whether because she's laying poorly, comes from stock susceptible to disease, or produces bees with a more aggressive personality—and remove her from her hive before combining. Next, lay a sheet of newspaper over the uppermost hive body of the hive that is still queenright. Then place the hive bodies containing the second queenless colony over the newspaper. Place your hive cover or lid on top of the hive, and come back in a week. By that time the bees should have chewed through the newspaper, slowly allowing their pheromones to mix, and now should be one bee family! You can consolidate the hive as necessary one week after combining.

Note that this method is for Langstroth hives. For a top bar colony, the same method can be used: simply drape the newspaper over the last top bar in the hive, and place the bars containing the queenless colony behind the newspaper. Return in a week to consolidate as necessary.

Is My Colony "Splittable"?

You will not be able to split every colony every year. Years with good rain and weather conducive to flower and nectar-producing plants will result in stronger colonies that can spare resources for a split, or colonies on the verge of swarming. Of course, other circumstances may result in weaker colonies that cannot contribute resources to a split. There are many reasons why a colony may not be strong enough for an increase, including a colony that went queenless or a queen that is failing or has run out of sperm. A colony experiencing disease or a high mite count will not have the population to contribute to a split. A colony that has already swarmed for the season may be lower in food and population. And of course, factors external to the colony can play a role as well: if the weather patterns did not provide adequate pollen and nectar to help support a growing colony, you likely will not be able to safely take resources for a split. So how do you know if your colony is "splittable?" This is where beekeeping becomes more of an art and less of a science. As you become a more experienced beekeeper, you'll become more accustomed to what constitutes a "strong" versus a "weak" colony, both of which are relative terms. Remember, all resources do not have to come from one colony. Examine each colony individually and determine what resources the colony could safely spare. All the frames in your brood nest need to be covered with bees, and if this is not the case for a colony, I do not recommend taking brood or bees from this colony. If you have a strong colony with wall-to-wall bees and brood, this colony may be a good candidate for sharing bees and brood. Remember that your split will require nutritional resources as well. You can feed a split sugar water, but it's always best to provide nectar and honey from existing colonies if possible. If you have assessed that a colony is low enough in stores that you are feeding it sugar water, you should not be taking food resources from this colony for a split. Once you've assessed each colony, calculate how many total frames of brood and bees you can safely take, and if it's not *at least* three frames of brood and bees and one frame of honey, nectar, and pollen, you probably do not have enough resources in your apiary to make a split at this time.

CHAPTER 10

Parasites, Pathogens, and Pests

Honey bees, like all organisms, can succumb to disease, predators, and pests that prey on their colonies. While learning about pests and diseases of honey bee colonies can be a little upsetting and worrisome, unfortunately they are realities of beekeeping. It is, therefore, very important to understand the pests and diseases that may inhibit your colonies and to understand how to prevent and manage them.

Working as a superorganism has many benefits to a honey bee colony, but unfortunately living in close quarters with thousands of other organisms means disease can spread rapidly throughout a colony. Most of these diseases spread through the collective feeding of the larvae, which means pathogens tend to infect a honey bee at a very young age. Also, honey bees have a strong tendency to rob from other colonies. This means that if a colony that is suffering from a disease has grown too weak to defend itself or failed, the disease can be easily spread to other robbing bees in the area.

Studying pests and diseases of honey bees can be an overwhelming topic, and the information in this chapter is dense. While it is important for you to read this chapter at least once in full, this was written as a reference guide to refer back to as needed. I've provided an overview of the most important and common pests and diseases. Research is ongoing, and what we understand about diseases and available remedies and treatments changes frequently, so ongoing reading and learning will be important in your journey.

Integrated Pest Management

Before we jump in, I want to first provide an overview of *integrated pest management*. Integrated pest management (IPM) is a strategy for managing pests and their damage through a combination of techniques that minimize the risk to people, property, resources, and the environment. In an IPM strategy, pesticides are used *only* after monitoring indicates their need and only according to established guidelines. The most effective way to manage pests is through a multi-pronged approach. IPM strategies are divided into categories and presented in the shape of a pyramid, with the least toxic strategies found at the bottom of the pyramid and the most toxic found at the top. Proponents of IPM encourage focusing as much of your energy on the bottom of the IPM pyramid as possible. This means enacting methods and practices that focus on prevention and that are less toxic to you, the environment, and the bees. Approaches for managing pests are grouped into four categories.

- **Cultural control:** The base of the pyramid is cultural practices, which are aimed at preventing pests with zero toxicity. These may include sanitation and hygienic practices, habitat modification, and use of resistant stock and genetics. An example of cultural control would be choosing native disease- and pest-resistant plants for your garden or bees with mite-resistant genetics in your bees for your hives.
- **Physical and mechanical control:** The next level in the pyramid centers on using physical and mechanical control with zero toxicity. Physical control may include using heat, cold, light, humidity, and sound to control pests. Mechanical control may include traps, barriers, and other physical manners of controlling the pest. Examples include using netting to prevent birds from eating fruits and using mulch to prevent germination of plants considered undesirable weeds. For beekeepers, this might include freezing frames or bars removed from the hive before storing to discourage pests that feed on the contents of the wax comb.
- **Biological control:** Biological control entails using parasites and predators to control for pests. For example, gardeners may use ladybugs in the garden to control for aphids. While research on the subject is new, there are some promising studies on using fungi to control certain common honey bee parasites.

- **Chemical control:** The uppermost portion of the pyramid is the use of treatments to control for pests. Chemical control options typically are grouped into two categories. The first includes "biorational" pesticides, which are organic or natural pesticides that have lower toxicity to the non-target organisms and the environment. The second category, found at the very top of the pyramid and therefore one that should be used with the most care, are more conventional, and generally more toxic, synthetic pesticides and antibiotics. Many beekeepers use antibiotic or pesticide treatments inside the hive to control pests, but these should always be used with care to minimize their negative effects on the bees.

As we get into the different pests and diseases of honey bee colonies, I will share different remedies from across the IPM pyramid.

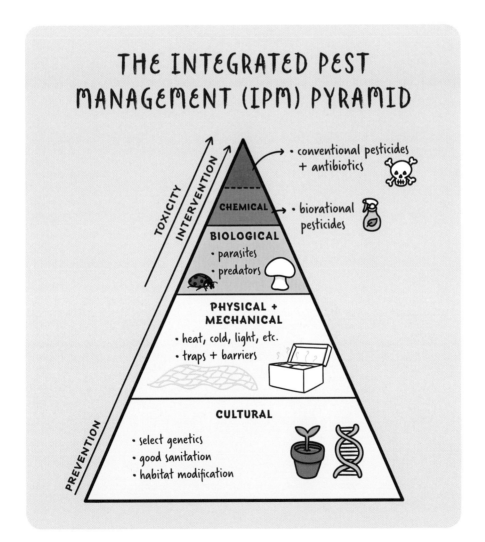

The Parasites: Varroa Mites

Varroa mites, or *Varroa destructor*, are tiny external parasites of honey bees, and the most devastating of all the pests affecting honey bees worldwide. The mite was first a pest of *Apis cerana* (the Asian honey bee), which has evolved over time to have a natural defense against varroa mites and no longer experiences the catastrophic colony losses that continue to befall *A. mellifera* (western honey bees). This is in part because varroa mites have been found in western honey bee colonies only relatively recently. The first varroa mites were found on imported honey bees in Hong Kong, China, and Singapore in 1963. They next appeared in Europe and South America in the 1970s, and they appeared in the United States in an apiary in Florida in 1987. They have since quickly spread to every continent in the world where honey bees live. In fact, as I write this in 2022, news is just breaking that varroa mites have been found for the first time in Australia.

IDENTIFICATION

Varroa mites are an external pest found on adult honey bees and on drone and worker pupae. All the varroa present outside of brood cells are female adult varroa, and they are oval-shaped, reddish-brown mites that are 1.5 mm wide. They look similar to tiny little ticks, and, similar to ticks, they spread diseases through their bites.

Unfortunately, although varroa mites are visible to the human eye, it's impossible to assess the number of varroa in a hive with a visual inspection only. Despite the difficulty of observing varroa mites inside the hive, there are signs that a colony is suffering from the effects of a high mite count, which beekeepers call parasitic mite syndrome. These signs include small puncture holes in

Varroa mite on bee larva.
© 2009 by beeinformed.org.

Brood suffering from parasitic mite syndrome.
© 2013 by Rob Snyder, beeinformed.org.

Testing for Varroa

There are several tests you can perform to assess the infestation level of varroa in a colony. The Honey Bee Health Coalition recommends testing each of your colonies four times per year.* Two of these tests involve taking a sample of ½ cup of nurse bees, roughly three hundred bees, from your colony and placing them in a jar with either alcohol or powdered sugar. The powdered sugar and alcohol help dislodge the mites from the bees, so you can count and assess the mite levels in the sample and extrapolate the results to the rest of the colony. Unlike the alcohol test, powdered sugar will not kill the honey bees. To conduct the sample, you will need a mason jar with a #8 screen mesh or hardware cloth affixed under the lid. A #8 gauge means the mites can be shaken through the screen but the bees will stay inside the jar. You also will need a white plate, a small plastic tub, a collection device measuring ½ cup, and either rubbing alcohol or powdered sugar. You can use an actual measuring cup or a plastic cup with ½ cup line marked on the side of the cup.

To collect the sample, locate a frame of open brood in a colony that is completely covered with bees, but be certain the frame does not contain the queen bee. Open brood is necessary because the test is meant to count the number of mites on the recently emerged nurse bees. Collect a sample by vigorously and quickly shaking the brood frame downward over the tub so that the bees fall into the tub. Use the collection device to collect ½ cup of bees, and pour the bees into the mason jar. If using the powdered sugar test, the jar should already have two tablespoons of powdered sugar in the jar. If you are using the alcohol test, add enough rubbing alcohol to completely cover the bees. Next, vigorously shake the jar for one minute and then let the jar sit in the shade for three to five minutes. Then, pour the alcohol or shake the sugar out of the jar and onto the white plate. The now dislodged mites will also leave the jar, and the white background of the plate will allow you to easily count the mites. If using powdered sugar, it's helpful to pour a bit of water over the sugar to dissolve it and make it easier to count the mites.

You can now extrapolate the results to the rest of the colony. For example, if you find three mites among the three hundred bees, the colony has a 1 percent infestation rate. If you find nine mites, the colony has a 3 percent infestation rate. As with all things honey bees, you will find a wide range of recommendations of when to take action, but a common recommendation is to intervene when the infestation rate is around 3 percent or higher. You can find videos of how to perform a sugar shake test on the Two Hives Honey YouTube channel, and more information on both tests on the Honey Bee Health Coalition's website.

*The Honey Bee Health Coalition is a coalition of beekeepers, farmers, researchers, government agencies, conservation groups, and consumer brands that provides education resources to farmers and beekeepers.

the capped brood, or the bees uncapping the pupae altogether, a phenomenon called bald brood. This occurs when the bees uncap the cells to remove pupae infested with varroa mites. This trait, known as varroa sensitive hygiene, or VSH, is a highly desirable trait in honey bee colonies. Other signs of a more advanced varroa infestation include a swift and significant drop in the honey bee population, a spotty brood pattern, and pupae that appear to have been chewed down. Varroa mites are vectors of viruses such as acute bee paralysis virus and deformed wing virus, and signs of these viruses are often found in colonies with high varroa mite counts.

THREAT TO THE COLONY

Varroa mites are one of the biggest threats to European honey bees today, and they are definitely the most destructive of all the pests. Varroa feed on the fat bodies of the developing brood at a critical time for the development of the bees and therefore have a devastating effect on the bees' long-term health. (The fat bodies of a honey bee are similar to the liver of mammals.) Brood infested with varroa often perish, but those that do survive and emerge as adults also experience negative effects from the viruses associated with varroa: this can include a shortened life span, an impaired ability to fly, and a weakened immune system. This weakened immune system can make the honey bees more susceptible to the viruses associated with mites, such as deformed wing virus, and these viruses can often result in the ultimate failure of the colony.

Of course, in small numbers this may not have a significant impact on the colony, but when the number of affected brood cells and emerging bees becomes relatively high, this will affect the entire colony's ability to rear and care for brood, forage for food, and defend the hive. Honey bee colonies often see their highest infestation toward the end of the summer months, and if action is not taken, they may go into the fall severely weakened and unable to prepare for and survive the winter months.

LIFE CYCLE

Varroa mites are parasitic, which means they require a host to survive and reproduce. Varroa mites can reproduce only in brood cells and must feed on either brood or adult honey bees. Adult female varroa mites will enter brood cells just before a larva is capped and will lay two to five eggs after the brood is capped. After hatching, the mites will move through the larval stage and become adults, all within the sealed brood cell. Even mating happens in the brood cell. The male varroa, which do not have mouth parts and cannot feed, will die shortly after mating. The female varroa mites emerge from the cell with the adult honey bees, and the cycle repeats. Because multiple mites can emerge from each cell, the mites can outpace the number of honey bees very quickly during a bad infestation. Because drones have a development cycle that is three days longer than that of worker brood, varroa are about eleven times more likely to lay eggs in drone brood.[1]

After emerging with the adult bee, the varroa mites will attach themselves between the hard scales of an adult bee's abdomen, feeding on the fat bodies. Because they attach to the adults, they can easily be carried out of the hive and spread to other colonies through activities like swarming, robbing, and foraging. When brood are not present, varroa cannot reproduce, although the life span of an adult varroa mite can be two to three months. As a result, climates that allow for year-round brood may find a higher incidence of varroa than climates that allow for a natural break in brood due to cooler temperatures.

PREVENTION

After mites were found in the United States in the 1980s, beekeepers who did not treat their hives with miticides experienced huge losses in colony numbers. But some colonies did survive. Researchers asked beekeepers to donate queens from these survivor colonies for study. Researchers at the USDA Honey Bee Breeding, Genetics, and Physiology Lab in Baton Rouge discovered a line of genes, now known as the VSH trait, that encourages the worker bees to recognize mite-infested brood and remove the pupae from the cells, killing any developing mites. This trait is not associated with one subspecies of honey bee but rather is a trait that can be selectively bred. Therefore, the most important varroa prevention tool is ensuring that your queen bee possesses this heritable VSH trait. You will know when your bees are exhibiting this behavior because you will see the previously mentioned bald brood: pupae that are uncapped in the hive. This is why researching the breeder who produces your nuc or package of bees is critical. Some beekeepers are also proponents of finding their own "survivor stock"—that is, catching feral swarms. The thought goes that if a colony can survive in the wild without beekeeper treatments and is strong enough to swarm, they possess some level of mite resistance. Just remember that a swarm may very well be of feral survival stock, or it may be a swarm from a neighboring beekeeper that does treat for mites.

MANAGEMENT

When thinking about varroa management, it's important to understand and accept that complete eradication of pests is rarely attainable and will likely never be possible with varroa mites. Looking back at the United States' reaction to varroa, we see that more than a decade of heavy toxic chemical treatments did not result in eradication but rather produced mites resistant to some of those treatments. Therefore, learning to manage your colonies while accepting that some level of varroa mites will be present is critical. As with all pests and diseases, I encourage beekeepers to focus as much of their energy on the bottom of the IPM pyramid as possible. This means enacting methods and practices that focus on prevention and are less toxic to you, the environment, and the bees. No silver bullet exists, so be sure to look for ways to employ different methods for preventing and managing varroa. I'll review each level of the IPM pyramid and, for each, provide some options for managing varroa. Know that some of these may not be right for you and your bees but are provided for your information.

Cultural

Examples of cultural practices to control for mites include the following.

Select desirable genetics: As previously discussed, make sure you select queens with the VSH trait to start your apiary. If you are relying on swarms to catch "survivor stock," keep in mind that not all swarms are feral; they may be from a nearby managed apiary.

Site hives with good sun exposure: Mites are temperature sensitive, so situating hives to receive at least one-half day of full sun and good ventilation can help with mite loads.

Physical and Mechanical

While there are no traditional trap methods for controlling varroa, there are a few other mechanical practices you may decide to employ.

Cull drone brood: Some beekeepers choose to cull, or remove, drone brood from their hives. Since varroa are eleven times more likely to reproduce in drone cells, this can be an effective method of removing developing varroa. Special plastic frames are available that encourage bees to build drone cells on the frame. Once the queen has laid drone eggs in the cell and they have been capped by the bees, a beekeeper can remove the frame and also any developing mites. If you try this method, be mindful to watch this frame carefully and remove it before the drones (and any varroa) emerge. But you don't need a special frame to cull drones: simply cut the cells out of the frame with a knife or a hive tool. Pro tip: chickens love to eat drone brood! I do some culling in yards where the number of drones in a colony may be "excessive," but I don't cull all drones from my apiaries. Although developing drones can be a magnet for varroa mites, drones also are critical to ensure that virgin queens can mate!

Use screened bottom boards: Some beekeepers choose screened bottom boards so that if the bees groom mites off of one another, the mites fall through the screen board out of the hive. This is a great example to demonstrate that not all tools are right for all beekeepers. I do not use screened bottom boards in my hives, as in my climate honey bees do a better job of regulating colony temperature with solid bottom boards. However, some screened bottom board designs have an insert so that the screened bottom can become a solid bottom board as needed.

Comb cycle: Comb cycling is a hygienic process that can help remove built-up pesticides and pathogens in the hive, reducing exposure to substances that may weaken the colony and make it more susceptible to parasites. This involves removing two of the oldest, darkest frames or bars of comb each year. This is best to do at your first post-winter hive inspection, when a colony is small and therefore there is plenty of comb that is not in use and that can be easily removed.

Give brood breaks: Because varroa mites are reliant on honey bee brood to mate, a period when there is no brood in the hive will interrupt the reproduction cycle of the mites. This is what we call breaking the brood cycle, and it can be done a number of ways. One way is to kill the queen and then, ten to fourteen days later, install a VSH queen from a breeder. Make certain to cut down all the developing queen cells before installing the new queen. Otherwise the colony will not accept the new queen, instead continuing to rear their own virgin queens. You can also allow the colony to rear a new queen, but if you are concerned about the colony's resilience against varroa mites, replacing the genetics in the colony with that of a VSH queen may help with future varroa management. Another way to introduce a brood break is to place your queen in a queen cage in the colony for several weeks, preventing her from laying eggs. However, this method also does not go the extra step of introducing new genetics into the colony.

Try thermal treatments: Research has found that slightly raising the temperature of the brood nest can kill varroa mites.[2] This is a fairly new option to beekeepers, and contraptions are available that will raise the temperature of the brood nest for several hours to a level that stresses the mites but that the honey bees can manage.

Biological

While I am not aware of any commercially available biological methods for control of varroa, research in this area is ongoing, and there are several promising early findings. In 2021, researchers at Washington State University successfully bred a strain of fungus that effectively kills mites but leaves honey bees unharmed.[3] Similarly, the fungal expert Paul Stamets has been working to develop a mushroom extract that boosts honey bee resistance to many of the viruses commonly transmitted by varroa.[4] While these treatments are not currently available to hobbyist beekeepers, they represent a new direction in varroa control, and they may become available to hobbyists in the coming years.

Chemical

A number of chemical controls exist, from less toxic treatments created from essential oils to a more toxic class of miticides. It's important to note that in the United States, mites have developed resistance to a number of chemical treatments due to overuse. This is why it's important that if you decide to use any treatments, no matter how organic, you employ them as a true treatment and not a preventative measure. Refrain from treating all colonies based on a calendar schedule; rather, use a treatment on a colony only when monitoring has uncovered a mite load that exceeds your acceptable threshold. Also, a number of these treatments cannot be used when honey intended for harvest is on the hive, because the treatment would make it unsafe to consume. Others can't be used with brood present or cannot be used at certain outdoor temperatures. The Honey Bee Health Coalition has a mite management tool that acts as a sort of decision tree.[5] You can input the characteristics of your colony (if there is brood present, for example), and it will share a number of options from across the IPM pyramid.

Two Hives Honey is a treatment-free operation, but that does not mean we ignore the presence of varroa or that we don't have to battle varroa from time to

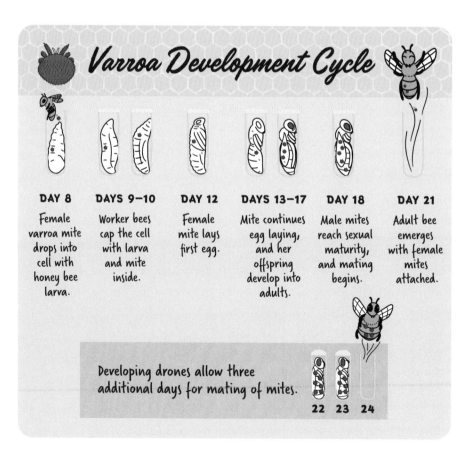

Varroa Development Cycle

DAY 8
Female varroa mite drops into cell with honey bee larva.

DAYS 9–10
Worker bees cap the cell with larva and mite inside.

DAY 12
Female mite lays first egg.

DAYS 13–17
Mite continues egg laying, and her offspring develop into adults.

DAY 18
Male mites reach sexual maturity, and mating begins.

DAY 21
Adult bee emerges with female mites attached.

Developing drones allow three additional days for mating of mites.

22 23 24

time. We rely heavily on genetics and brood breaks to control varroa in our apiaries and have been very successful in avoiding chemical treatments as a result.

It's also important to note the incredible amount of ongoing research on varroa and its effect on honey bees. Just in the few years ahead of me writing this chapter, scientists discovered that varroa actually feed not on the blood of honey bees, as once thought, but rather on the fat bodies. Further, at the writing of this book Australia just had its first varroa sighting on the continent. What we know about managing, treating, and preventing varroa changes relatively frequently, so it's important to find ways to stay abreast of new developments.

The Pathogens

WHAT DOES HEALTHY BROOD LOOK LIKE?

Before we get into different brood pathogens and diseases you may encounter, it's important to know the signs of healthy brood so that you can identify when disease is afflicting a colony. The young larvae should be moist and swimming in a healthy amount of royal jelly. They should not appear dry. Older larvae should be in a "C" shape, pearly white, and plump. They should not appear discolored or shriveled, or look melted in the cell. Healthy pupae should be sealed without holes or perforations, and the cells should not appear greasy. It's important to note, however, that brood comb gets very dark with use very quickly. Newly built comb in a hive is white as snow, but as generations of brood cycle through the comb, the worker bees will line the cells with propolis and remnants of the pupae's cocoons will be left behind, quickly darkening the comb and causing it to be less malleable. Brood comb will quickly become black in color. Many beekeepers mistake this for some disease of the hive, but this is a natural side effect of colony activities.

There are more pathogens than are included here, but those listed are either the most common or the most detrimental.

EUROPEAN FOULBROOD

Cause and transmission: European foulbrood (EFB) is a disease of the brood caused by the bacterium *Melissococcus plutonius*. The disease affects the larvae in a hive and is transmitted by the ingestion of contaminated food. The bacteria will multiply in the midgut of the larvae and compete with the larvae for food, eventually killing the larvae before they reach the pupal stage. EFB is associated with nutritional deficiencies and other colony stressors. I most often see EFB in colonies in the early springtime, when populations are rapidly increasing but the nectar flow has yet to begin.

Identification: A colony infested with EFB will have larvae that look twisted

Brood suffering from European foulbrood.
© 2005–2013 by Rob Snyder, beeinformed.org.

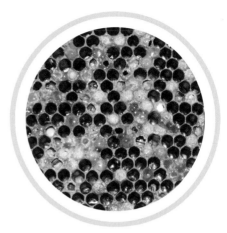

Brood suffering from American foulbrood.
© 2005–2013 by Rob Snyder, beeinformed.org.

in the cell. Rather than pearly white in color, they will instead turn slightly yellow, and eventually brown, as the larvae die. After they perish, the larvae will be easily removable. The brood pattern also will be spotty in nature. It's important to understand how to identify EFB because it can be mistaken for another disease called American foulbrood (AFB), which is far more serious. Keep reading to learn more about AFB. Diagnostic kits can be purchased at beekeeping supply shops to help diagnose EFB.

Prevention: Minimizing stressors to colonies can help prevent EFB. Ensure that a colony has proper ventilation by elevating hives and making sure the apiary gets three to four hours of sun, minimum, each day. Be sure to understand the nutritional needs of a colony, and supplement these needs if they are not being met naturally by the ecology.

Actions to consider: If EFB is a result of nutritional deficiencies, the disease often clears once a nectar flow begins. For this reason, feeding a ratio of 1:1 sugar to water to the diseased colony weekly until the nectar flow begins can also help. Refer back to chapter 7 to see best practices for feeding. Remember to adjust the amount you feed to the size of the colony—never give a colony more than a gallon per week, and never give a colony more than it can consume in a week's time. In colonies where EFB is present alongside other stressors, such as a high varroa count, you will have to address these issues as well. Requeening a colony with a queen bred for hygienic qualities can help, and the resulting break in the brood cycle can also aid the colony. We don't use any chemical treatments or antibiotics in our apiaries, but know that a veterinarian can also prescribe antibiotics to treat EFB.[6]

Cause and transmission: Similar to EFB, American foulbrood is a bacterial disease of honey bee brood caused by the spore-forming bacteria *Paenibacillus larvae*, which multiplies in the gut of the infected bee. It is spread by feeding contaminated food to developing larvae. The bacteria that causes AFB can survive for dozens of years and is extremely contagious. These spores also can be spread through tools and equipment or through honey and pollen from infected colonies. Once a colony with AFB fails, it can then be spread via robbing bees stealing honey from the infected colony. Unlike EFB, which usually affects colonies already suffering from nutritional deficiencies or other colony stressors, AFB can affect even the strongest of colonies.

Identification: Capped brood affected with AFB will appear discolored and have a greasy appearance, with sunken cappings. As the brood perishes it will form tough scales in the bottom of the beeswax cells, which will be hard to remove. You can perform the "ropiness" test by putting a matchstick into cells showing symptoms. If withdrawing the matchstick produces a long, brown, ropy thread, it may be an indicator that the colony is infected with AFB. However, this test can be subjective and may not give a clear indication of whether the infection is EFB or AFB. Remember that in EFB, the brood will die before the pupal stage, and with AFB, brood usually perishes after the cells are capped. Pupae that have perished from AFB may have their tongues facing upward in the cell. Also, colonies affected with AFB usually have a very strong and offensive odor of decay that is hard to miss. You can purchase an AFB diagnostic kit at large beekeeping supply shops.

Prevention: For all brood diseases, keeping honey bees whose genetics allow for hygienic behavior is key. This means the honey bees will detect, uncap, and remove any infected brood. Requeening colonies that do not exhibit this behavior with queens that are bred for this trait is important. For all brood diseases, the beekeeper should also practice hygienic beekeeping practices: do not share frames from colonies whose brood appears to be diseased, and if you suspect AFB in your colonies, do not use any tools and equipment that have come into contact with the colony until you confirm the diagnosis. Never feed your bees honey from hives that you have not inspected and cannot confirm are free from disease. This includes any honey purchased from a grocery store, which has been found to often be contaminated with EFB and AFB. In January 2023, the US Department of Agriculture granted a conditional license to a biotech company to produce and sell a vaccine for AFB—the first vaccine ever approved for insects.[7] The vaccine will initially be available only to select commercial beekeepers. Though the effectiveness of the vaccine is still unclear, it is a very interesting development in the fight against a devastating disease that affects our honey bees.

Actions to consider: Thankfully, though AFB was rampant in the early part of the twentieth century, it is no longer the primary cause of honey bee deaths in the United States thanks to more hygienic practices, awareness, and destruction of infected colonies. If you suspect you have AFB, please contact your local department of agriculture or state apiary inspector immediately. Other brood diseases can easily be mistaken for AFB, and your state officials can help diagnose your colony. AFB is not curable, spreads quickly, and can kill even the strongest of colonies. Many states have laws requiring your colonies infected with AFB to be destroyed and hives to be burned.

NOSEMA

Cause and transmission: Nosema is a disease affecting adult honey bees caused by a spore-producing fungus that enters a honey bee through infected food or water. Two different fungal parasites cause nosema in honey bees: *Nosema apis* and *Nosema ceranae*.[8] These parasites are very similar in their effect on the colony and in prevention steps. Once the fungal spores from each parasite enter the midgut of the honey bee, they germinate and begin reproduction. They absorb the nutrients from the bee's gut, and the honey bee is weakened and now susceptible to other infections. The spores will pass through the bee's digestive system and can contaminate food and water sources through fecal waste. *N. apis* infections usually appear in the early spring after a winter confinement, when honey bees are unable to make cleansing flights to relieve themselves and therefore defecate in the hive. Nosema is very common in temperate climates.

Identification: Nosema can be difficult to identify because adults suffering from the disease show no outward change in appearance and the signs of the infection, such as a declining population, reduced brood production, and bees crawling at the hive entrance, are not unique to a nosema infection. Honey bees suffering from nosema are unable to produce royal jelly and will move to the foraging function more quickly, ignoring brood-rearing duties. Bees will also avoid consuming sugar syrup. Nosema is often accompanied by dysentery, which is identified by yellow or brown streaks along the front of the hive near the entrance. Please note, however, that just because a colony has dysentery does not mean it also has nosema, just as all colonies with nosema will not have dysentery. Laboratory testing of adult bees is the only way to truly diagnose nosema. If you suspect nosema may be affecting your colony, contact your state office responsible for beekeeping for information on labs that are available to test for nosema and for more information on collecting your sample.

Prevention: As with all brood diseases, it is important to keep colonies strong and limit stressors, including nutritional deficiencies. Because nosema is often found when honey bees are unable to take cleansing flights, make sure your hives are in a place that maximizes the amount of sunlight and ventilation they receive and minimizes moisture during the cooler months. If you are in a temperate climate, ensure your hive entrances face south, if possible, to ensure maximum sunlight in the winter months. Hygiene practices such as recycling out old brood comb is a best practice for helping to prevent nosema.

Actions to consider: Often nosema will clear up on its own as temperatures rise, honey bees are able to leave the hive to relieve themselves, and older infected bees die off. However, when monitoring determines that the intensity has reached a certain threshold, an antibiotic called fumagillin is approved for use against nosema in the United States. The antibiotic is fed to honey bees in syrup and will prevent the spores from reproducing but does not kill the spores. You can buy the antibiotic at beekeeping supply stores and online.

CHALKBROOD

Cause and transmission: Chalkbrood is a disease of the brood caused by the fungal spores *Ascosphaera apis*. It spreads when the spores are ingested by larvae. The spores start to absorb nutrients from the larvae, causing the larvae to die of starvation. Chalkbrood is most common just after the winter dearth, when temperatures can still be cool at night but the colony has begun to rear spring bees and the brood nest starts rapidly growing. When temperatures drop at night, the clustering bees cannot adequately keep all the developing brood warm, and the chilled brood are then susceptible to chalkbrood. The brood usually dies soon after the bees cap the cells.

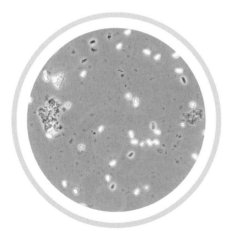

Nosema spores under a microscope.
© 2022 by beeinformed.org.

Chalkbrood. Conall.

Identification: Chalkbrood is identified by hard, mummified-looking dead larvae that turn white, then gray, and finally black. You may find the mummified larvae at the entrance of the hive as the bees work to clean the cells. Chalkbrood is named as such because the dead larvae resemble pieces of broken chalk.

Prevention: Similar to nosema, you can help prevent chalkbrood by placing hives in well-ventilated areas that receive good sunlight. This will help keep the hive warm and dry during the winter months. Combine weak colonies before the winter dearth to help ensure strong colonies.

Actions to consider: Chalkbrood usually will not cause huge losses of bees in a colony and often will clear up as temperatures rise. Consolidate space in weak colonies to ensure that the colony can best maintain hive temperature. Requeening with a hygienic queen is recommended for colonies suffering from chalkbrood.

SACBROOD VIRUS

Cause and transmission: Sacbrood most often affects brood but also can be found in adult honey bees. It is caused by a virus from a family of viruses known as *Iflaviridae*. The virus is transferred to young larvae through brood food. Nurse bees can be infected by dead larvae as they clean cells of diseased larvae, unintentionally ingesting the virus during these duties. Sacbrood may be found more frequently when other stressors are affecting the colony, such as poor nutrition.

Identification: Infected larvae die after capping but before pupating and can be seen with their heads facing upward once the bees uncap the cell. The larvae will change color from white to yellow and then brown, and they will appear as a fluid-filled sac. Similar to nosema, infected adult bees will ignore brood-rearing duties and move to the foraging functions more quickly.

Prevention: As with all brood diseases, focus on keeping strong healthy colonies, limit stressors such as poor nutrition, and replace poorly laying queens.

Actions to consider: Sacbrood generally is not a cause for concern, but if the disease is rampant, requeening with a queen with hygienic properties can help.

OTHER VIRUSES

There are thirty-six different viruses that affect honey bee colonies. Several of these viruses rely on the presence of parasites.

Varroa mites feeding on honey bees are the vectors of two of these, deformed wing virus (DWV) and acute bee paralysis virus (ABPV). Both viruses can affect both brood and adult honey bees. DWV causes deformity in adult honey bees. Mites will obtain the virus from infected bees and transmit it to other colonies while feeding on the fat bodies of the honey bees. The affected bees' wings may

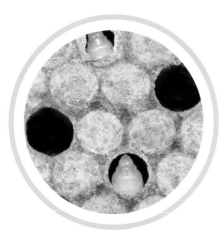

Brood suffering from sacbrood.
© 2013 by Rob Snyder, beeinformed.org.

Adult small hive beetle.
Courtesy James D. Ellis.

appear shrunken and singed, as if they had been burned with a candle, and their abdomens may appear shortened. However, only colonies with varroa infestations show signs of the viral infection.

ABPV can be spread via feeding, but transmission of ABPV via food sharing alone generally isn't enough for symptoms to appear. As with DWV, varroa mites play an important role, as they transfer the virus directly into the hemolymph of the adult bees and brood. Hemolymph is a fluid similar to that of blood in vertebrates. Often no sign of ABPV is found before the colony fails, but heavily infected bees can show signs of paralysis. When a colony's population dips very suddenly, ABPV as a result of a varroa mite infestation may be to blame.

Because varroa mites are necessary for a colony to show symptoms of DWV and ABPV, taking all precautions to help control for varroa mites is the most effective prevention for each. Once a colony is heavily infested, even varroa treatments may not be enough to save the colony.

The Pests

SMALL HIVE BEETLES

Small hive beetles (SHB) are a pest native to parts of Africa that can now be found in beekeeping communities around the world. They were first discovered in the United States in the mid-90s in a Florida apiary.

Identification: SHB adults are 5 mm to 7 mm long. Young adults are brown but turn black as they grow older and their exoskeletons harden. The ability to identify the larvae is critical, as the larvae do the most direct damage to a colony. The larvae are cream color and are 9.5 mm long, with spines all along their bodies and three sets of developed legs near their heads. Adult beetles can fly long distances.

Threat to the colony: SHB thrive in warm, humid conditions, which is why they are endemic to Australia and the southeastern part of the United States. The SHB is what is known as an opportunistic, or secondary, pest; this means they take advantage of an otherwise weakened colony. If a colony is stressed by other factors and cannot adequately guard and protect itself, SHB can move in quickly and overwhelm the colony. A strong colony usually can control the SHB population. Though the bees cannot remove the beetles from the hive nor sting through their exoskeletons, they will chase the beetles into corners and crevices of the hives, often using propolis to "jail" the intruders.

If left unchecked, the larvae will burrow through the combs, consuming eggs, pollen, and honey. The real damage occurs when a large number of SHB are present and are able to defecate in the honey, which releases a yeast that causes the honey to ferment. The honey and combs will then take on a slimy appearance and a foul odor, which is why beekeepers call this phenomenon "getting slimed." The honey is unfit for human or bee consumption. When the infestation reaches this level, the queen may cease laying eggs and the colony might abscond, abandoning the hive altogether.

If you live in a cooler climate, SHB may not be a concern, as they only thrive in warmer and humid climates. For those areas affected, as we are here in Texas, I equate the presence of SHB in hives to cockroaches in city apartments or mice in country homes: they are ubiquitous, and every single hive you have will have at least a few. Also similar to cockroaches, each SHB sighting usually means there are many more elsewhere that you aren't seeing.

Life cycle: Adult female SHB lay dozens of eggs in crevices of the hive or

Small hive beetle larva.
Courtesy USGS Bee Inventory and Monitoring Lab.

directly into the combs. Within two to four days, the eggs will hatch into larvae, which will tunnel through the hive, feeding on pollen, honey, and bee brood, for roughly two weeks. Next, they must leave the colony to pupate in the soil. They will leave the hive and burrow approximately four inches into the soil, building a sort of cocoon to continue their development inside. The emergence of the adult SHB depends upon moist soil and warm temperatures and may take as little as two weeks or more than a month. (This is why the threat of SHB is higher during particularly rainy seasons.) Once an adult beetle emerges, the cycle starts over, as she is able to fly to infect new colonies, locating them based on their odor.

Prevention: The best defenses against SHB are to ensure that hives have a minimum of three to four hours of full sunlight every day and to focus on understanding how to keep your colonies strong and healthy. Good sun exposure can dry the soil and help disrupt the pupation of the beetles, and strong colonies can almost always control SHB without intervention. A colony weakened by another event is most susceptible to the destruction they can cause. Of course, if you see SHB during your inspections, I encourage you to smash as many as you see with your hive tool. However, overworking your hives just to try to take care of the problem with physical force interferes with the honey bees' natural ability and processes to control for the beetles.

Further, minimize the areas where SHB can hide in your hive: remove any debris and burr comb from your bottom boards, and replace warped or cracked equipment. Never give your colony more space than it can adequately protect, and watch weak colonies carefully: do not leave them with excessive amounts of honey that they cannot guard. Share that honey with another, stronger colony until the colony can grow stronger to protect itself. If you have a dead-out, or failed, colony, remove the hive and any honey immediately to prevent an infestation that can infect nearby colonies.

When harvesting honey, be sure to extract the honey within forty-eight hours. Unguarded honey supers are a breeding ground for small hive beetles, and therefore honey should be processed immediately or stored in the freezer. If you are not ready to harvest the honey, leave it with a strong colony until you can process it to ensure that it is not slimed by SHB.

Management: You can find myriad traps and systems for SHB control. Many require the use of some bait, like apple cider vinegar, or some substance to drown the SHB, such as mineral oil. In our apiaries, we focus on keeping our colonies strong and then use a product called Beetle Bee-Gone, which is a cloth-like sheet that traps the SHB. You place the sheet in the hive and the bees will chew at it, which causes the fabric to felt into a ball. The bees will chase the SHB into the cloth, and the beetles' legs get tangled in the fabric. Once the sheet is full of SHB, simply remove it and replace with a new sheet. Some clever

The Mammalian Predators

There are a few mammalian predators that may trouble your bees. Skunks and raccoons may approach at night, scratching on the hive to elicit attention, then snag a few bees for their supper as they come out to investigate. Raccoons may also enter the hives by taking off the top lids in an effort to steal a snack. We have both skunks and raccoons at our Honey Ranch, and a few key actions in the apiary keep them at bay. First, put Langstroth hives up on stands eight to sixteen inches tall. Both of these critters have tender underbellies, so if they have to stand up on their hind legs and expose that area to reach a hive, the honey bees will take care of the rest. Second, place heavy rocks or bricks on the top of the hives to help deter raccoons from opening them. Mice on the hunt for warm places to nest may be an issue in the winter months. With honey bees in cluster, they can't adequately guard the entrance to keep them out. You can find mouse guards at beekeeping supply stores. My anecdotal experience, however, tells me that if you live in an area with mild winters, the threat of mice may be low. I have an apiary that is full of field mice but have never used a mouse guard and have not found a single mouse in an active hive with bees in almost ten years. (I know the mice are there because dozens of them nest underneath my hive stands and in empty equipment on-site.) I suspect that because we have many winter days with temperatures close to or above 50°F, the bees can do a great job of keeping the mice at bay. However, in climates where bees stay clustered for weeks or months on end, you may find mice to be a bigger concern.

Bears, another mammalian predator, are a bit trickier to defeat. Of course, we don't have any bears in Central Texas, so my experience is limited to what I hear from other beekeepers. But I have seen the damage that a bear can do, and it is eye-opening how quickly a bear can rip through an apiary, pulling hives apart in an instant. If you're curious, do a quick internet search for images of hives destroyed by bears. The damage is astounding! Interestingly, contrary to what all the cartoon bears may tell us, bears aren't after just the honey: they also seek out the brood and bees for protein. If you have bears in your area, I strongly recommend installing an electric fence to keep your apiary safe.

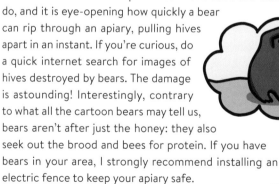

beekeepers also have discovered that the unscented disposable cloths used to mop floors are made of the same material and can be used. However, I find that these are thicker and may trap bees as well. Some beekeepers have had some luck treating the soil with nematodes or diatomaceous earth.

Remember that eradication is not possible, and keeping bees in an area where SHB is present requires acceptance that you will have some level of SHB in your hives. Focus on understanding how to keep colonies strong, remove other stressors, and give hives at least a half day of full sun, and you likely will never have to deal with a truly problematic infestation of SHB.

WAX MOTHS

Two types of wax moths infest honey bee colonies: the greater and the lesser wax moth. They are similar in life cycle, damage to the colony, and prevention and management, so I will discuss them together.

Identification: The adult greater and lesser wax moth are similar in appearance and grayish in color, though the greater wax moth is larger, at 13 mm to 19 mm long, than the lesser wax moth, at 10 mm to 13 mm long. Many beekeepers confuse wax moth larvae and SHB larvae: wax moth larvae are much larger than SHB larvae, and they have a darker-colored head. SHB larvae are 9.5 mm long, while the lesser wax moth larvae are 13 mm and the greater wax moth larvae are 28 mm. SHB larvae have just three sets of legs near the head, while wax moth larvae have three larger sets of legs near their head and smaller legs along the length of their body as well. Wax moth larvae also tend to be a dingy, grayish color, while SHB larvae are more cream colored.

The first indicator of moths may not be the actual adult or larvae. As the larvae tunnel through the beeswax comb, they leave behind a trail of white silk

Adult wax moth. © 2005–2013 by
Rob Snyder, beeinformed.org.

Wax moth larvae. © 2005–2012 by
Rob Snyder, beeinformed.org.

that looks like webbing. Also, if a wax moth larva is tunneling through brood comb, you may see what is called bald brood, or uncapped pupae, where the bees are uncapping the cells to try to access and remove the larvae. Another sign of wax moths is white cocoons attached to the woodenware in your hive.

Threat to the colony: Similar to SHB, wax moths thrive in warmer conditions and are considered an opportunistic pest. Bees in strong colonies can manage populations very effectively, but if a colony has excess brood comb that it cannot adequately patrol, wax moths will take advantage. The larvae will tunnel through the hive, eating beeswax, pollen, and bits leftover from the cocoons of emerging honey bee brood. The greatest threat of wax moths is the damage they do to beeswax comb once it is removed from the hive. A strong colony is very effective at preventing wax moth infestations, but wax moths can destroy any unguarded comb very quickly, leaving behind massive amounts of silk and defecation.

Life cycle: Wax moths' life cycle is similar to that of honey bees: egg, larva, pupa, then adult. The speed at which the moth moves through each stage of the life cycle is temperature dependent, with optimum temperatures for development at 84°F to 92°F. After mating, a female wax moth begins laying eggs immediately, laying between 50 and 150 eggs in crevices of the hive. The eggs hatch in three to thirty days, with development occurring more quickly at higher temperatures. After hatching, larvae move through the comb, feeding and destroying the beeswax. The larval stage is the only stage of the development cycle in which the wax moth consumes food. The larval stage lasts anywhere from twenty-eight days to six months before the larvae spin a cocoon and pupation begins. At this stage, the larvae chew into the woodenware of the hive, damaging the equipment, and spin a cocoon from silk. The pupal stage lasts as little as three days in warmer climates or up to two months in cooler temperatures before an adult emerges, starting the cycle once again.

Prevention: Because wax moths are secondary pests, the best defense is the same as with SHB: focus on keeping your colony strong and eliminating any outside stressors. Manage your colony's space to make sure colonies do not have more space than they can guard. Remove dead-outs from the yard immediately.

Finally, protect your brood comb once it is removed from a hive. There are a number of reasons why you might have brood comb that is not residing in any hive: perhaps you had a dead-out or needed to consolidate a weak colony. We often have lots of extra brood comb after we do consolidation of hives heading into our winter dearth. Generally wax moth damage isn't a concern during months when temperatures are regularly below 50°F, but any other time of year it is important to take precautions to protect any comb. Remember that bees require a lot of energy and carbohydrates to build beeswax comb. Preserving your comb

Lessons from a Beekeeper

MUHAHA

Many new beekeepers spend a lot of time worrying and fretting about opportunistic pests like SHB and wax moths. These pests aren't the cause of a colony's demise; rather, they are a symptom. Something else has almost always weakened the colony, and it is now unable to adequately guard the space. When folks tell me, "Wax moths killed my colony," I know they missed something else along the way. Claiming that wax moths killed your colony is akin to saying that maggots killed a dead animal. The issue is that if you continue to believe that wax moths killed a colony, you are missing an opportunity to diagnose what went wrong so you don't miss the signs again. Perhaps the colony went queenless and wasn't able to requeen itself, and then it slowly diminished over time. Or maybe the colony has a high incidence of brood disease or a more virulent pest such as varroa mites that is weakening the adult population. At the first sign of either of the opportunistic pests in any large numbers, start assessing the strength of the colony and how you can improve its numbers. Keep in mind that in warm climates where small hive beetles are endemic, such as Texas, every hive is going to have some small hive beetles during the warmer months. But a strong colony is really good at defending itself. All of your frames in your brood nest should be covered in bees—if they are not, either consolidate or begin to share frames of brood from other hives to help build your population. And remember that these pests do become a major concern when you are removing old drawn comb or honey from a hive and the bees can no longer guard the comb. Be sure to extract or freeze any honey within forty-eight hours of pulling it from a hive to keep small hive beetles from ruining your crop, and freeze and properly store old brood comb to protect it from wax moth damage.

to give to colonies later will help colonies grow more quickly during the growing seasons and store more honey during the nectar flow. It is important to note that generally only comb that contained brood or has stored pollen is at risk—if the comb held only honey, it is probably safe from wax moths. You can tell whether comb had brood stored because it will be darker in color where the bees lined the cells with propolis.

It's easy to protect brood comb from wax moths. First, freeze any brood comb pulled from a hive for forty-eight hours to kill any wax moth eggs. You can either keep the comb in the freezer until you're ready to put it back on a hive or store the combs by double bagging them or putting them in an airtight container. If they are not stored properly, wax moths will infest the combs. Before storing, be sure to defrost the combs and ensure that they are dry; otherwise, mold will form on the comb.

Management: If you do find wax moths in any large numbers in your colony, it is almost assuredly because the colony is not strong enough to guard the space it has been allotted. Consolidate the hive, and if the colony is weak, share capped brood from another colony if possible. With infected combs, cut out and dispose of the affected areas of the comb, then freeze what remains and store properly as described previously. If the entire beeswax comb is ruined, you can cut out the webbing and remove any cocoons and still use the woodenware—disposing of the hive components is not necessary.

Though pests and diseases vary in their potential harm to a colony and the available prevention and remedies, here are a few best practices that can help better ensure that a colony has the best chance to stay disease free and pest free.

1. **Ensure good nutrition.** It is critical to ensure that a colony has access to proper nutrition, either from the natural environment or supplemented if the environment cannot meet the colony's needs. A well-fed colony is better equipped to ward off many brood diseases than a colony stressed by poor nutrition.

2. **Give proper ventilation.** Ensure that colonies have proper ventilation. This means placing a colony where it has several hours of full sun each day, at a minimum, particularly if you live in a cooler climate. Avoid damp, dark areas for apiaries.

3. **Practice comb recycling.** Periodic recycling of the oldest brood comb can help with the hygiene of a colony. Buildup of pesticides and disease spores in the comb over time can affect the health of a colony. Some beekeepers may note the date that a frame or top bar is put into rotation in a hive by writing the approximate date that the colony draws the comb on the top of the frame or bar. Then they may remove these after anywhere from three to five years. Another method is to simply inspect your colony on your first post-winter hive checks and remove one or two of the oldest, darkest frames or bars from the brood nest that are not currently in use by the bees. You accomplish this by deciding what brood frames or bars would ideally be recycled during the winterizing of a colony. Place frames or top bars on the outside of the brood nest in the hopes that the resulting smaller colony after the winter season will not be using the old comb. Then simply replace with new frames or top bars.

4. **Reduce stressors.** In general, try to limit the stressors on a colony. Leave bees with plenty of honey for the dearth periods. Do not inspect colonies when temperatures are cold enough that the bees are close to clustering. Place hives in an area with an abundance of varied pollen sources and nectar.

5. **Connect with your local apiary inspection service.** Most states in the United States have one or more inspection services. Connect with the representative there to learn more about their services and help they can provide.

6. **Continuing your learning and research.** It is important to understand that what we understand about honey bee pests and diseases is constantly evolving. For example, two developments during the writing of this book required edits to this chapter before it was printed: the appearance of varroa in Australia and the new vaccine approved for AFB. Continue to stay abreast of new developments in this area.

CHAPTER 11

Harvesting from the Hive

A honey bee hive is a treasure chest of products that provides an ample number of health benefits. Harvesting is an exciting time for a beekeeper and an opportunity to experience the fruits of your and your bees' labor. In fact, the most common question I get, after "How many times have you been stung?" is "How much honey will I get from one hive?"

Unfortunately, this question is impossible to answer without a lot more context and information. First of all, let's set some expectations: I discourage beekeepers from expecting a harvest their first year. Unless you bought a fully established colony, your nuc or package of bees needs time to grow, build beeswax comb, and store honey for the winter dearth. In the first year, simply focus on learning more about the colony and ensuring that it is strong and heavy with honey before winter.

Before a beekeeper even begins thinking about a harvest, they need to first ensure that the colony has produced enough honey to sustain itself. Most climates provide access to nectar for only short periods of the year, which means the bees will depend on those honey stores the rest of the year to ensure they don't starve. Only if a colony produces above and beyond what it needs to make it through the dearth should you consider a harvest.

So how much honey will you get from one hive? Well, you very well might get none. If a colony is weak during the nectar flow, the bees don't have access to enough nectar, or the weather patterns don't cooperate, it may be challenging for the bees to make enough to feed themselves, much less you. Of course, a strong colony with a bountiful nectar flow can produce several dozen pounds of extra honey in a year. Some years provide the right conditions for a heavy nectar flow, and some years it's leaner. This is the unfortunate reality of working in agriculture.

Given all that, remember that if you keep bees for the experience, you will never be disappointed. If your primary objective is to produce honey, I can guarantee a lot of disappointment ahead. Fortunately, putting the bees and the health of the colony first involves the same actions needed to maximize honey production. Win for all!

The Weather Factor

Much of what happens in beekeeping is not within your control. Beekeepers rely on undomesticated animals to produce the final product, and weather can make or break a honey crop in a season. If this is your first foray into agriculture, you may be a little shocked at how much the weather report suddenly matters! Even experienced beekeepers are at the mercy of the colony's wishes and the whims of the weather. No matter what a beekeeper does, a colony may swarm anyway and take a large percentage of the worker population and honey stores just before a nectar flow. An especially rainy season can ruin a crop: bees can't forage in the rain, and the rain washes the nectar from the blooms or knocks the blooms off the plants. The flowers will replenish, but several weeks of nonstop rain at the wrong time can severely limit the amount of nectar the bees can gather. Similarly, drought can inhibit a plant's ability to grow and produce a bloom with nectar. Get ready to be constantly dissatisfied with the local weather forecasters!

Of course, not all factors are beyond the control of a beekeeper. An experienced beekeeper's hives are likely to produce more honey than those of a less experienced beekeeper. Keeping colonies strong, healthy, and disease-free goes a long way toward improving honey production, and these skills come with time and experience. Economies of scale are present in strong colonies: two colonies of twenty-five thousand bees each will make less total honey than one colony of fifty thousand bees. This is why swarm control ahead of the nectar flow is critical to making honey!

The Nectar Flow

Nectar flow, or honey flow, is a term used to describe the time when bees have access to so much nectar that they can convert it and store it as honey. In most climates, honey bees do not have access to nectar most of the year. This means that when the nectar is flowing, they turn on their hoarding tendencies, converting the nectar into honey to store for the coming months or years. Bees often have access to small amounts of nectar other times of the year, but once the flow is on, they can finally stop living "hand to mouth" and start storing away the excess. (This tendency makes me think of my Nana, spending hours in the kitchen canning the bounty of tomatoes and more from my grandad's spring garden!)

A Colony on a Nectar Flow

The nectar flow is a critical time for a colony, and the colony has been preparing for it for months, growing in size and building its foraging population. So how do we know our bees are on a flow, and what can we do to help ensure that they are able to store away as much honey as possible?

Talking to and learning from experienced local beekeepers can be a great way to learn about your local nectar flows, as the flows in most areas occur around the same time each year. Your area may have one main flow and several smaller flows throughout the year. For example, in Central Texas the largest nectar flow starts in mid-May and lasts through June, and a second, smaller flow usually lasts from late September into October.

There are several signs that indicate whether bees are on a nectar flow. Seeing a plethora of nectar-producing flowers can be one clue. The "busy-ness" of your hive entrance can be another. But both of these are just clues and cannot tell you definitively whether your colony is on a flow. Sometimes Mother Nature tricks the flowers, and they may be blooming but not yet producing nectar. And a hive that is being robbed by other bees could be mistaken for a colony on a flow. I've heard some beekeepers report that they can even tell that bees are bringing in nectar because their bellies are swollen. If that's the case, they have better eyesight than me because that's not a skill set I possess!

The only way to know for sure whether a colony is on a flow is to inspect the hive. A colony is on a flow when the bees are building beautiful new white beeswax comb. Remember that bees cannot build the beeswax comb necessary to hold new honey without the carbohydrates found in nectar. Once the nectar flow is on, the bees will take advantage of the excess carbohydrates and build additional sheets of comb to store the incoming nectar. If the nectar flow is strong enough, they will continue curing the nectar into honey. Look for gorgeous comb, as white as fresh snow, above and around the brood nest. A strong nectar flow is almost impossible to miss, as long as you are inspecting your hives regularly. If one colony is on a flow, don't take for granted that all colonies in the apiary also will be on a flow. Remember that the only way to know for sure is to inspect the colony! Small, sickly colonies may not be able to bring in nectar in large amounts, even if it's available.

What to Do during a Nectar Flow

The nectar flow is a critical time for a colony, and what you do or don't do can dictate not only whether your hives produce any excess honey to harvest, but, more importantly, determine whether your bees have enough honey to feed themselves until the next nectar flow.

Once a colony is on a nectar flow, make sure the bees have additional space in the hive to store all the new nectar and eventual honey. In a Langstroth hive, this means adding hive bodies. A hive body added to a Langstroth hive in preparation for a nectar flow is called a super, and this act is referred to as "supering" a colony. In a top bar hive, this may mean moving the follower board so that the bees have access to more top bars for storing honey. Though you've learned you should never give a colony more space than it can adequately guard, the eve of a nectar flow is the one time of year I recommend providing a colony more space than it may immediately need. Nectar flows are critical times for a colony, and they may not last more than a few weeks. I'd rather be two weeks early than one day late in adding space to a colony on the eve of a nectar flow! Our spring nectar flow usually starts in mid- to late May, so by the first week of May I make sure my strongest colonies have a super ready to go on the hive. A strong colony can guard this additional space without any difficulties. Smaller colonies will be unable to make as much honey, so less space can be afforded. If the bees don't get on as strong of a flow as expected, you can reduce the space later. Other actions may include the following:

- Open up or remove entrance reducers altogether. This is not a requirement, and it is important to note that in the wild, bees actually often choose cavities with smaller entrances. However, a smaller entrance will inhibit the foragers' ability to move quickly in and out of the colony on a strong flow.
- Add entrances. I live in a hot climate that can be fairly humid when the nectar flow is on. Adding an upper entrance can help hot, wet air escape and assist the bees with dehydrating the nectar. Plus, an upper entrance can provide a more direct path to the nectar storage area for your foragers if they choose to use the entrance. You can make an upper entrance by drilling

a hole in the super or using a shim or stick between the hive bodies. In our yards we simply offset the uppermost super slightly, creating a small gap on one corner of the hive. In a top bar hive, open additional entrances if the hive has them.

- Stay out of brood nests. Unless a colony has a concern I have been monitoring, such as a colony that was previously diseased or queenless, I stay out of the hive for the three- to five-week nectar flow period. Hive inspections are disruptive, and I don't want any interference during a nectar flow. Once the nectar flow is completed, I will do a thorough inspection of the hive once again. However, a beginning beekeeper may wisely choose to continue inspections during a flow to assist with better understanding of this process.

Assessing the Harvest

Once the nectar flow is completed, inspect the hives to determine whether a colony has harvestable honey. Before you ask, "How much honey can I take?" you must first ask, "How much honey do my bees need?" The answer to this question will depend on the strength of the colony, the strength of nectar flows in your area, and the length of time until the next nectar flow. Stronger colonies with more bees require more food. Colonies in areas with more frequent and stronger nectar flows won't need as much honey to survive the dearths than those in areas with weaker nectar flows. Depending on where you live, the colonies may need anywhere from thirty to ninety pounds of honey. In Central Texas, we have shorter winters and earlier opportunities in the year for nectar flows, so we can leave less honey. Beekeepers in the northern part of the United States have longer, harsher winters, and nectar may not be available again until July or later. They therefore must leave more honey for their colonies. Chat with local experienced beekeepers for their recommendations. Your own experience collected from one year to the next will help inform your decisions as well. This is yet another reason to keep very detailed notes!

Once you know how much honey to leave, just a bit of math will determine how much honey you can safely harvest. In a Langstroth hive, a medium frame of honey, when completely full on both sides, will contain four to six pounds of honey. A deep frame, when completely full, will contain six to eight pounds of honey. This isn't an exact science but can help provide a good estimate of how much honey a colony has stored. If using a top bar or other type of hive, use a scale to get a better idea of how much a full frame or bar holds. Add up all the honey and nectar stores in your hive, then subtract the amount needed for the bees. The total remaining is the *maximum* amount you should consider harvesting.

Shoo, Bees!

Once you determine how much honey can be harvested, you must remove the bees from these combs. When selecting frames or top bars, look for ones whose cells are at least 80 percent capped. Taking frames or top bars with uncapped cells means you may be taking nectar that has not yet fully cured, and the nectar will eventually ferment. Do not take any frames or top bars with any stages of brood. And, of course, make sure to use the math you learned in the last section and only take surplus honey!

Several options are available to safely remove the bees from the combs:

- **Shake and brush:** The shake and brush method works well if there are only a few combs to harvest. Using a vigorous, short downward motion, shake off as many of the bees as possible. If you are using top bar hives, skip this step, as the combs in a top bar hive can be fragile and fall off the bar easily. The shake will achieve a 90 percent solution, and then you can use a bee brush to gently brush off any remaining bees. The brush method is the only option available for top bar users.

- **Bee escape:** A bee escape is a clever contraption that has a sort of one-way exit and can be used in Langstroth hives. First, assemble all your frames in as few boxes as possible. Next, place the bee escape in between the top box that will stay with the hive and the supers that you plan to harvest. Close up the hive. Over the course of a few hours, the bees in the supers will use the bee escape exit to go down into the hive and out to forage again, but they won't be able to find their way back up into the supers through the bee escape. Several hours later, the supers should be free of bees. Please keep in mind that though the bee escape is tricky to navigate, it's not impossible, and honey bees are smart critters. If you leave the supers and bee escape on for more than twenty-four hours, the bees may find their way back into the supers.

- **Fume board:** A fume board is my preferred method because it requires one visit to the bee yard and the bees are removed from an entire super all at once. A fume board is a contraption with a panel that absorbs heat on one side and felt or cloth on the other side. Fume boards also require the use of fumigant, a liquid that has an odor that tends to repel bees. You can find several on the market, including a few organic options. To use a fume board, first choose a day that is warm and sunny for the task. The fumigant doesn't work well on cool, cloudy days. In the bee yard, place the fume board in full sun with the solar panel facing up so it can begin heating up. Next, assemble all your harvestable frames in as few supers as possible on top of the hives. Be sure to use the shake technique we described before placing the frames in the super to be fumigated. Next, spray a light layer of fumigant across the felt side of the board, making sure to get coverage from end to end. Place the board, felt side down, on top of the super to be fumigated. Do not use an inner cover or top cover anywhere on the hive during this process. Wait three to five minutes for the bees to run away from the offensive smell, then pull the fume board off to expose the now bee-less super! Be sure to not leave the fumigant on for more than a few minutes, as this will cause all the bees in the hive to vacate as a result! I find fume boards work best when fumigating just one super at a time.

HOW TO USE A FUME BOARD

1 Place solar-absorbent side in direct sunlight.

↑ outside absorbs heat

2 Assemble your frames in as few supers as possible over a hive.

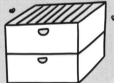

(Use a bee brush to brush the bees off, if possible.)

3 Flip the fume board over and spray the inside thoroughly.

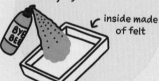

← inside made of felt

4 Place the board, felt-side down, on the super to be fumigated.

5 Wait 3-5 minutes while the bees flee.

6 Unveil the bee-less super!

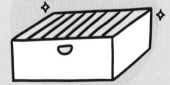

Your Honey versus Grocery Store Honey

You may be a little surprised by how different your honey appears and tastes from that sold in big grocery store chains. Honey produced by small-scale beekeepers differs from big brand names in several ways.

First of all, your honey will be minimally processed and will therefore be raw honey. Raw honey is honey that has not been heated at high temperatures and filtered through very fine filtration systems. Honey sold by big brands undergoes this pasteurization process because it produces a honey that will never crystallize. Unfortunately, this pasteurization process not only negatively affects the flavor but also removes most of the health benefits of the honey. On the other hand, your raw honey will likely crystallize. Honey crystallization is a natural process and does not mean the honey has spoiled! Rather, it confirms that the honey is raw and retains all the great health benefits naturally found in honey.

Furthermore, the honey sold in grocery stores by bigger name brands is almost always what is known as blended honey. Blending honey is a process by which honey from many beekeepers across different areas (and likely even different countries) has been mixed together. However, different flowers produce nectars with different attributes, which results in a rainbow of colors, flavors, and textures of honey. When honey is harvested in small batches, the honey retains those unique properties. For example, our hives just a few miles apart in Austin can produce vastly different types of honey depending on what flowers are blooming in that neighborhood, and our spring harvest is vastly different from our fall harvest. When big honey packers blend all the honeys they buy together, the honey loses those unique attributes and develops a more generic flavor profile and color.

Your family, friends, and customers also may be surprised at the vibrancy of the color and flavor of your honey. Be sure to educate them on why your honey is better than those big brands, and that crystallization has no effect on the quality or the freshness of the honey.

Extraction and Bottling

Now that the honey has been safely pulled from the hives, let's discuss options for extracting the honey from the frames. The method you choose will depend on the products you ultimately wish to produce, the resources available, and the amount of money you are willing to dedicate to harvesting tools.

Before beginning, keep in mind these few tips:

- Cover the workspace. Honey is sticky. I'm sure you are aware, but it's hard to imagine how challenging it is to clean up honey until you do it on a large scale.
- Work indoors. Honey cannot be extracted outdoors or in the bee yards. The presence of honey will cause a robbing frenzy, drawing in every bee in a three-mile radius. Work in a space indoors where the bees cannot access the honey.
- Work in a warm room. Honey doesn't flow well when it's cold.
- Do not harvest until you are ready to complete the job. All sorts of pests, including small hive beetles and ants, are ready to take advantage of the now unguarded honey. Do not pull the honey from the hives unless you are ready to begin work extracting within the next forty-eight hours or are prepared to store the frames in the freezer.

Following are just a few methods for extracting honey from the frames and top bars.

CRUSH AND STRAIN

This is a perfect method for a small-scale beekeeper or a beekeeper with top bar hives. The process is in the name: crush and strain! Using your hands or other tools, crush the beeswax comb in order to break the cells open, allowing the honey to flow down with the help of gravity. First cut the comb from the bar or frame and then use clean hands or even a potato masher to crush the comb. Make sure to break open every cell. Do this over some straining mechanism, which can be a colander or cheesecloth stretched over a bowl or bucket, or a cheesecloth bag. Allow the comb to strain for twenty-four hours. The benefits of crush and strain is that it requires little to no financial investment, and if you clean the wax, you can filter and render the fresh beeswax to make candles, lip balm, and other body products. The downside to crush and strain is that it destroys the beeswax comb that could otherwise be placed back in the hive for the bees to refill with nectar.

EXTRACTION

Extraction is a method in which beekeepers use a centrifuge to sling the honey from the frames of a hive. An extractor can be used only with Langstroth-style frames whose combs are secured with foundation and/or wires running through the frame. I also have had some luck using foundationless Langstroth frames in an extractor if the bees have secured the comb on at least three sides.

Extractors come in all sizes, dictated by the number and size of frames they can hold, and can come in manual or electric models. Extractors also come in two types: radial and tangential. Radial extractors hold the frames like a wheel holds its spokes, and both sides of the frame are extracted at the same time. With a tangential extractor, the sides of the comb run the length of the extractor and honey is extracted only from one side of a frame at a time. To complete the extraction, flip the frames and run the extractor again.

Before frames can be extracted, the beeswax caps must be removed from the hexagon cells. Several tools are available at beekeeping supply stores to assist with this part of the process. I recommend buying several different types and trying them out. At minimum I recommend buying a tool known as an uncapping fork. A serrated bread knife found in most kitchens can also be a great uncapping tool. Make sure to remove the cap on all the cells—you will not be able to extract honey from any cells whose caps remain intact. Be sure to complete this uncapping process over some sort of vessel to catch and drain the honey off the caps. Uncapping tubs or tables can be found at beekeeping supply stores. A DIY option is as easy as placing a queen excluder over a bucket.

Once the caps have been removed from both sides of the frames, place the frames in the extractor and begin spinning the extractor, starting slowly. If you have a larger extractor and will not be using all the spaces available for frames, be sure to load the frames in a way to ensure that the extractor is well balanced. An unbalanced extractor acts much like an unbalanced washing machine! The extractor type and speed will dictate how long you will need to run the extractor before the frames are drained. Read the extractor manual for more instructions. If you are using a tangential extractor, be sure to flip the frames and run the extractor again so the honey is extracted from both sides of the frames.

After extraction is complete, the honey must be strained to remove any remaining bits of beeswax. I recommend purchasing a basket honey strainer from a beekeeping supply store, a tool consisting of two strainers that fit inside one another. The top strainer has a wider mesh gauge and is meant to catch the larger pieces of wax and debris, and a finer mesh sits below, straining out the finer bits of wax. Place the basket over a bucket, open the extractor's honey gate, and watch it go! Be mindful not to abandon the extractor during this process or your strainer can overflow, sending your precious honey all over the ground.

Lessons from a Beekeeper

As you learned in the last chapter, SHB and wax moths are low on my worry list because they are opportunistic pests. Focus on keeping your colonies otherwise healthy and strong, and these pests won't be much of a bother. However, these do need to be top of mind when removing honey or brood comb from a hive for storage. Once the honey and brood comb have been removed, the bees can no longer protect it from these pests, and you must be extra cautious to protect them. During a harvest one season, we experienced what is known as a bumper crop of honey. A bumper crop is a harvest that far exceeds normal yield. It sounds like a great problem to have, but the tricky thing about bumper crops is that the farmer usually doesn't have the labor, processes, or machinery to effectively and efficiently process the extra yield. Two Hives was just a few years old, and we were still figuring out key processes when we were overwhelmed with honey from this bumper crop. At that time I was still mostly solely responsible for extracting honey. As a team we would bring in honey from the apiaries and stack the supers, and I would work my way through the stack, starting at the top. The other beekeepers would come in, pick up the extracted supers to take back to the yards, and leave more supers for me to extract. By the time I made my way to the bottom of the stack at the end of harvesting, I realized the bottom three supers had come in several weeks prior—I had not managed my extracting to ensure that I was making my way to the bottom of the stack in a timely manner. I found three supers of honey, totally slimed by SHB. To add insult to injury, the supers were mesquite honey, a customer favorite and our best seller. Over one hundred pounds of beautiful mesquite honey was ruined. The lesson here: before you harvest the honey, make sure you are prepared to actually extract the honey from the comb in the next forty-eight hours. If you're not ready, leave the honey with the honey bees so they can safeguard it for you! (Also, when you get a bumper crop of honey, find extra hands immediately!)

HARVESTING METHODS

1 CRUSH AND STRAIN

MASHING TOOLS

potato masher

hands

STRAINING TOOLS

cheese cloth

strainer

RECEPTACLES

bowl

bucket

2 EXTRACTION

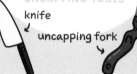

extractor

strainer

UNCAPPING TOOLS

knife

uncapping fork

3 CHUNK/COMB HONEY

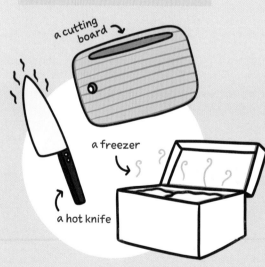

a cutting board

a freezer

a hot knife

For bottling honey, use a **honey gate** to make filling jars easier!

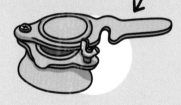

The greatest benefit of an extractor is that it preserves the hexagon beeswax cells for the bees to reuse later. It also is a pretty quick process for removing the honey from the frames, though the size of your extractor and whether it is manual or electric will determine how long the process will take. The drawback to extractors is that even small manual models are not cheap. You also won't get as much beeswax to use for rendering as you would with the crush and strain method.

HARVESTING COMB HONEY

Of course, the rawest and, in my opinion, best way to eat honey is right out of the comb! Comb honey is very easy to produce if you are using top bars or Langstroth hives without any foundation in the frames. Once the honey is capped, simply lay the frame or top bar on a cutting board and use a knife to cut the comb honey into squares or whatever shape you desire. An important tip: using a hot knife will give the comb honey a beautifully clean cut and will avoid crushing the comb as the knife cuts through it. And the comb honey *must* be placed in the freezer for forty-eight hours to kill any wax moth eggs that may be in the comb. I find it's easiest to do this after you package the comb honey. To produce chunk honey, simply place a piece of comb honey in the jar and fill the rest of the vessel with liquid honey.

No matter what method you choose, look for a honey gate at your local beekeeping supply store. This is a valve that when installed on a plastic food-grade bucket allows for easy bottling of honey into smaller jars.

After extracting the honey from the frames, frames and bars that are wet with sticky honey are left. Place these back inside your hive, and within twenty-four hours the colony will clean up the frames and bars, ensuring that not even a drop goes to waste. All the other tools, knives, and more can be placed in the apiary for the bees to clean up as well. Keep in mind that placing anything with honey outside will cause a robbing frenzy, so be sure to place all items at least thirty feet from the hives. If you live in a more urban area on a smaller lot, avoid using this cleanup method, as the robbing frenzy can frighten neighbors!

Honey Harvesting Best Practices

Here are a few tips to help ensure maximum honey production and make the harvesting process run more smoothly:

- Focus on keeping colonies strong and healthy, and the honey will follow. Strong, large populations will always make the most honey, so if you focus on practices that keep colonies healthy, you are also doing what's best for honey production.

- Don't forget that pollen does not equal honey. Too many experienced bee-keepers see bees bringing in pollen and presume their colony must also be on a nectar flow. Just because flowers produce pollen does not mean they are providing nectar as well. Remember: inspecting your hives is the only sure way to confirm whether they are on a nectar flow.

- Find a work-around if the honey supers are too heavy to lift. If you are concerned with your ability to move entire supers for harvesting and ex-traction, fill your boxes with only as many frames as you can safely carry. However, fill any empty spots with empty frames in the super until the box is full—carrying around supers with just a few frames can cause the frames to bounce onto the ground, ruining the hard work of your bees!

- Keep frames and boxes of honey covered as you work in the bee yards. Robbing is easy to incite, and the scene can become intense. Robbing bees can cause huge losses of honey and bees in a hive. I use corrugated plastic boards to keep supers covered while we work. These are easy to find during election season, as they are often used as political yard signs.

- Be smart about your tools. Lots of gadgets can be found in most kitchens to simplify the process and minimize the investment required. However, there are a few specific beekeeping tools that I strongly recommend. One of these is a food-safe bucket outfitted with a honey gate for ease of bottling. If you are using an extractor, I also recommend a two-part honey strainer and an uncapping fork.

Harvesting and Using Propolis

Propolis is one of my favorite non-honey products to harvest and use from the hive. Studies have found that propolis is an antibacterial, antiviral, and anti-inflammatory agent.[1] It's frequently used in products intended to fight colds and sore throats as well as promote healing from minor cuts or burns. It also is used in skincare formulations.

It's easy to harvest propolis. Propolis traps can be found at beekeeping supply stores, but I've found the easiest and best time to harvest is just after the winter dearth. There will be lots of areas in the hive where the bees have added propolis to keep out the cold, often around the entrance and in gaps between hive bodies. Simply use a hive tool to scrape the propolis that is sealing up the entrance of the hive, place it in a container, and store it at room temperature until you have enough to make propolis products.

Many beekeepers harvest propolis to make formulations for colds, treatments for burns and wounds, and toothpastes and mouthwashes. Here are recipes for a tincture and a throat spray that we use in my household when we are feeling a little under the weather.

RECIPE

Propolis Tincture

INGREDIENTS
2 PARTS PROPOLIS
   MEASURED BY WEIGHT
9 PARTS CLEAR GRAIN
   ALCOHOL

I CHEESECLOTH
I MASON JAR
I DARK-COLORED
   DROPPER BOTTLE

DIRECTIONS
Mix the propolis and alcohol together in a canning jar. Store in a dark place, being sure to shake the mixture 2 to 3 times a day for I to 2 weeks. Strain through the cheesecloth and store in an opaque dropper bottle. The propolis tincture can be ingested internally. Take 4–5 drops under your tongue daily to support good health.

RECIPE

Propolis & Honey Throat Spray

YIELD: 5 TBSP

INGREDIENTS
2 TBSP PROPOLIS
   TINCTURE

2 TBSP RAW HONEY
I TBSP WARM WATER

DIRECTIONS
Mix all ingredients in a spray bottle and use to help soothe throats. You can also add other herbal ingredients to further the healing or help with the taste, such as elderberry, ginger, or peppermint.

Other Products of the Hive

Honey and propolis are far from the only products a colony can provide. Beekeepers can manage their hives to produce pollen, bee venom, beeswax, and even royal jelly. Plus, honey can be manipulated to produce new and fun products as well, including infused honey and creamed honey. Making infused honey is a process in which beekeepers use time, heat, or both to add flavors to different honeys. Making creamed honey is a process in which beekeepers control crystallization to produce a honey with a texture similar to that of smooth butter. This process involves using a seed honey, agitation, time, and a bit of heat to ensure that your honey takes on the smoothness of your seed honey.

Should I Sell My Honey?

In one word, yes! I always cringe when I see students giving away huge amounts of honey to family and friends. I understand the want and need to give away some honey as holiday gifts, but please be mindful of giving away large amounts of honey after the harvest. Doing so sends the message to the community that honey is easy to produce and has no value. If you've gone through the trouble of managing your colonies to get to a successful harvest, and then actually experienced a honey harvest, extraction, and cleanup, you know honey isn't easy to produce and jar. One honey bee will make $\frac{1}{12}$ teaspoon of honey in her entire life. A single weather pattern can ruin a honey crop, and I don't dare count the number of stings and the amount of my own sweat (and sometimes blood) that goes into producing, extracting, bottling, and selling a single jar of honey. Even if you aren't in the business of selling honey, posting on social media or inviting members of your club or church to purchase your honey helps promote the value of honey. And you should definitely charge more than what you see in the grocery store! Your honey is a far superior product. Grocery store honey often has been blended, heated, and heavily filtered, and perhaps even adulterated and cut with cheap sweeteners. Plus, beekeeping can be an expensive hobby, and charging for your honey helps pay for the cost of your time, tools, replacing colonies, or purchasing new hives.

Preparing Your Honey to Sell

Check with your local Department of Health for any laws pertaining to the bottling and selling of honey. Many areas have laws excluding honey from stricter regulations usually applied to making and selling food. For example, your state may have a law allowing the bottling of honey in a home kitchen as long as the

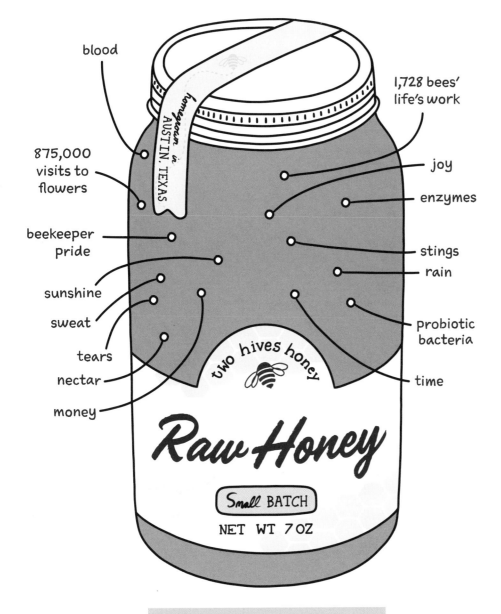

ANATOMY OF LOCAL HONEY

label includes a statement that the product wasn't produced in a state-inspected kitchen. However, these laws, often referred to as "cottage laws," may have additional restrictions on how the honey can be sold. Before engaging in any sort of commercial activity, make sure you understand the local regulations regarding the bottling, labeling, and selling of your honey.

The US Food and Drug Administration (FDA), the agency responsible for the regulation of food safety, has a number of labeling requirements that are important to understand and follow if you are selling honey in the United States. The FDA provided additional guidance in 2018 on the labeling of honey.[2] I've included a summary, but it is important that you research and follow the latest FDA guidance, or the regulations in your country if outside the United States, on the creation of your honey labels:

- If the food contains only honey, the food must be named "honey." Because it is a single-ingredient food, an ingredient statement is not required for the label. If any other sweeteners are added to the honey, it must be labeled accordingly.
- If any other flavors or ingredients are added, the name of the product must accurately describe the food with its characterizing flavor. For example, "lavender-infused honey." If any ingredients are added, an ingredient list must be included on the label.
- The net weight of the honey, excluding packaging, must be noted on the label. This is important, because the weight and volume of honey are not the same! A good practice is to measure the weight of the empty packaging (including the lid) and subtract that weight from the total weight of the packaging with honey to get the net weight.
- The label must include the name and the address of the manufacturer of the product. In this case, you would be considered the "manufacturer" of the honey.

Once you feel confident in the legal requirements of bottling and selling your honey, spend some time thinking about the brand of the product! Brainstorm a catchy name and logo that are meaningful to you and your beekeeping story. When I first got started, I made the leap and knew beekeeping and selling honey was going to be my full-time job, so I invested money on a professional graphic designer to help me create a logo and my labels. But if your aspirations are smaller, play around with one of the many free graphic designer tools you can find on the internet to come up with a logo and label. You can print labels at a local print shop (shopping local is always my preferred option if possible!), or you can find printing companies online.

Where to Sell Your Honey

If your beekeeping operation is pretty small, your first few harvests will likely be small enough that you won't have to reach too far beyond your social networks to sell the products. Many of my students have shared a few posts on social media and have sold most of their first few harvests to excited family and friends.

As your operation, and perhaps ambitions, grow, you may need to find new outlets to move your product. Many small operations get their start at local farmers markets. Research local markets and pay them a visit to scope out the existing honey market. Just because a local market already has a honey vendor doesn't mean you can't also sell there, though some markets do limit the number of types of vendors. Checking out the competition's offerings can help you as you think about the pricing of your products and even what products you may want to offer. Finding ways to differentiate your business and products will help earn more customers. For example, perhaps another beekeeper at a local market sells only jars of honey. You can choose to offer jars of honey but also think about offering chunk, infused, or creamed honeys. At Two Hives, our signature product is our comb honey. It's hard to find, and even fewer companies produce comb honey that does not contain beeswax foundation. Our customers know that we have the most tender comb honey without that extra waxy texture found in a lot of the other comb honey on the market.

Farmers markets can be a great option for small producers. They provide great exposure and allow you to actually talk to customers each week, which helps you learn more about their preferences, answer questions, and see their reactions to the products, pricing, and branding. But farmers markets are also labor intensive, and once you grow enough that you need to hire help, finding reliable weekend labor can be tricky. I took a different approach, looking to build a wholesale program that would allow me to focus on the beekeeping, agritourism, and beekeeping education programs. Leaning into our hyperlocal approach to beekeeping, I labeled all of my jars with the neighborhoods where the honey was produced and used that differentiating factor to get my honey into small markets, cheese shops, and restaurants. I later focused on building a direct-to-consumer approach, opening a small honey retail shop and growing my online sales.

Please remember that your local health department laws may limit how and where you sell if you choose to take advantage of local cottage laws. For example, taking advantage of a cottage law and bottling in a home kitchen (as opposed to a health-inspected facility) may mean you are not allowed to make online sales or wholesale retail sales. If you plan on making honey a true business, be sure to keep this in mind when you think about your growth and where you hope to sell.

CHAPTER 12

Beekeeping through the Seasons

Beekeeping is seasonal and regional. The needs and actions of a bee colony will change many times throughout the year and can look a bit different in apiaries even a few miles apart. A popular refrain is that "all beekeeping is local." This is often used as a reason to avoid giving general guidelines or best practices to beekeepers across different areas. While this refrain is true, the tactical aspects of beekeeping are exactly the same everywhere. Whether you are in North Carolina or North Dakota, the practices I have described in this book apply. This is because the needs of honey bees are the same everywhere.

So what is different? The climate.

A honey bee colony changes its behavior in response to the weather and the availability of food resources in the colony's ecosystem. The availability of these food resources, nectar and pollen, are a direct result of the local climate. Is the temperature right for flowers to bloom? Do they have enough sun, water, and nutrients in the soil? A flower that blooms and produces nectar in May in Texas may not bloom until July in South Dakota, or a northern ecosystem may not support that species at all. "All beekeeping is local" because beekeepers in different regions will have different resources available to their colonies at different times of the year.

A good beekeeper needs a deep understanding of the relationship between the flora and honey bees to make informed decisions about what actions to take in any given season. Many will give advice about beekeeping and relate it to the dates on a calendar. For example, you may hear a beekeeper in your area recommend harvesting honey in July. That can be helpful as a benchmark. However, if the weather doesn't cooperate and presents an atypical pattern, as is seen more and more as a result of climate change, your nectar flow may start later in the year and honey may not be ready to harvest until August!

Here are a few other examples of how unexpected weather patterns may affect colonies:

- A late-season hard freeze in an area can kill all the plants just emerging from the soil, which can mean little to no early-season pollen. The presence of pollen is one of the critical determining factors of when and how much brood a colony can produce, and the brood nest relies on an abundance of pollen for proper nutrition and development. A severe lack of pollen will cause a queen to slow or stop laying altogether, affecting the rate at which the brood nest can grow. And malnourished brood may produce bees that are more susceptible to disease.
- An unusually dry blooming season can mean blooming flowers that aren't actually producing any nectar, inhibiting a colony's ability to make honey.
- An unusually mild summer with lots of rain can produce a longer bloom period for nectar-producing plants in the fall, resulting in bees' ability to produce a larger crop of honey.
- Of course, a severely hot and dry summer can produce the opposite effect. Plants that have already germinated and are growing with the expectation that they will bloom in the fall can fail to produce blooms if they can't make it through an extreme summer. This is actually what is happening in Texas as I write this: all the goldenrod and sunflowers that would normally provide us a summer nectar flow are wilting in the heat, and I don't expect that they will bloom if we don't get rain soon.

Therefore, rather than focusing on seasons, it is more beneficial to discuss beekeeping in the context of the availability of resources. It's when we change the backdrop of the conversation that we learn that beekeeping really is all the same!

Of course, differences in climate are not insignificant. The climate will dictate your bees' foraging time, nectar and pollen availability, the ability of pests to thrive, and the insulation needs of your colony. For example, beekeepers in climates with long, harsh winters have to leave more honey on their hives than beekeepers in milder climates, but it's not because the bees or beekeeping is different. It's because the colonies in the colder climates have a much longer period without access to nectar and therefore must have larger stores of honey.

In this chapter, I will summarize what you can expect as resources ebb and flow, and actions you may need to take as a result. Refer back to previous chapters for a more detailed explanation of each of the management techniques you have already learned.

A quick note: I can't talk about weather without acknowledging climate change. Here in Texas we already have some pretty erratic weather patterns. With climate change, these periods of unusual weather and intense storms get worse each year. This means beekeepers have to constantly adapt to these patterns and alter what they expect of bees at any given point. Gone are the days when we were able to identify, almost to the day, when a nectar flow would start. Nectar-producing plants respond to weather patterns, and these increasingly unusual weather systems force plants and bees to change course as a result of weather and, in response, beekeepers to rethink colony management.

What's in a Season?

I grew up in Smyer, Texas, population four hundred. I distinctly remember being in elementary school and looking at our seasonal posters hanging on the wall, intended to teach us about spring, summer, fall, and winter. The fall poster always had brown leaves falling to the ground in the months of September and October, while the winter poster always had snowmen and snowballs in December and January. Fall in West Texas didn't bring fall leaves; it just brought days that were slightly cooler than the hot days of summer. Where I grew up, the average October high temperature is 84°F! (Plus, we don't really have trees for leaves to fall off of anyway out there.) I saw very few snowballs and even fewer snowmen in my eighteen years of growing up in my hometown. The opposite occurred when I moved to Washington, DC. I remember my frustration every year when we would get a snow or a freeze in April. It was April, for goodness' sake! Back in Texas we were already halfway to summer! The meaning of a season, I've realized since, is completely relative.

Climate affects the forage of bees and the colony's tendency to contract or expand. If you understand how these three work in tandem, then you can keep bees anywhere. You only have to understand the local climate and how it affects the availability of resources, and then you will know and understand how the bees will respond. So let's toss these notions of spring, summer, fall, and winter out the window and instead invent new seasons that more closely relate to resource availability and the resulting actions of the honey bees!

THE DEARTH SEASON

A dearth usually occurs during the harshest weather of the coldest or hottest months, as those are times when the flora is not as abundant. A dearth is a scarcity or lack of a resource. Beekeepers use this term to refer to times of the year when a colony has very little or no access to nectar. (A colony also can experience a pollen dearth, but given that the pollen needs of a colony are more limited than its nectar needs, the term is usually applied to nectar.) You may have one or more dearths in your area. Here in Central Texas we usually have two dearths: a summer dearth and a winter dearth. Of course, if you live in a tropical climate, you may have very short dearths, or perhaps none at all! Even here in Central Texas, with our relatively mild winters, we do have a dearth that begins around late October and lasts until February. Our summer dearth is the one that worries me most, because our incredibly hot and dry summers affect our bees' ability to adequately prepare for winter. During a dearth, a colony will rely on its stored food resources to survive the season.

What to Expect

During a dearth, a queen will slow down laying eggs or cease laying altogether. Your area may experience cold- or hot-weather dearths. During a cold-weather dearth, when temperatures fall below 50°F, a colony must work together to stay warm and avoid freezing to death. This is because at around 40°F or lower, a honey bee simply cannot operate its muscles. In these cold temperatures, the bees will cluster together in a tight ball, with bees crawling into empty cells to further tighten the cluster, vibrating and moving their flight muscles in order to generate heat. This action ensures that the middle of the cluster stays around 95°F, with the temperature lowering farther away from the center of the cluster.

Of course, a dearth can occur in extreme heat as well, and during this time the bees are working hard to keep the colony cool! Bees need a lot of water to cool the hive in the hottest months of the summer.

During any dearth, you should expect to see your colony size decrease over time as your queen slows or stops laying. During a hot-weather dearth, you may find honey bees trying to rob from one another in response to the lack of resources.

What to Do

The key to surviving a dearth is preparing for the dearth.

Cold-weather dearth: In the months leading up to a cold-weather dearth, inspect colonies to ensure they have plenty of honey to survive until the next nectar flow. What is "plenty" will vary wildly by area: in Central Texas we leave our colonies thirty to thirty-five pounds of honey, while beekeepers in the northern United States may leave double or even triple that amount. Check in with local beekeepers to understand how much food a strong colony will need. If you are worried about a colony not having enough food, either share honey from another hive or start feeding sugar syrup far in advance of the winter dearth. When feeding sugar syrup ahead of a cold-weather dearth, remember to mix a 2:1 ratio of sugar to water to make it easier for the bees to store, and increase the sugar content as colder temperatures grow closer.

Just ahead of the cold-weather dearth, you should take several actions to prepare the colony for the cold months:

1. **Create a honey "envelope" around the brood nest.** Honey is a great insulator, and the cluster will move as a unit in search of food, so honey should surround the brood nest.
2. **Reduce the hive entrance to help the colony retain as much heat as possible.** If you live in an area where mice may be common, using a mouse guard at the entrance can help prevent mice from moving into the hive over winter.
3. **Consolidate the hive down to only as much space as the colony needs to house itself.** Remember, from this point forward, a colony will begin shrinking. But it is important to have empty comb in the brood nest. This means the hive should not be packed full of honey throughout. Bees need empty comb to crawl into to keep the cluster tight and warm.
4. **Finally, colonies in very cold climates may require insulation to stay warm.** Here in the southern part of the United States this isn't necessary, but in areas with harsher winters, beekeepers often choose to insulate their hives.

Avoid checking a hive any time temperatures are below 50°F. Doing so can break the cluster. Bees can't move their muscles at low temperatures, so they may not be able to recluster and the colony can freeze to death! If you have to emergency feed colonies at these low temperatures, do it quickly and without disturbing the cluster, or wait for a warmer day when the bees won't be in cluster. Each time a hive is opened, its propolis seals are broken and the bees cannot rework the propolis to reseal a hive at lower temperatures.

In summary, start winter preparations in the months leading up to the dearth by ensuring the bees have plenty of food stores. After you winterize your colonies, relax! This is your off season and a time of rest for the beekeeper. It's also a great time to reflect on the year and mentally prepare for the next (and, of course, build beekeeping equipment!).

Hot-weather dearth: For areas that are prone to intense hot-weather dearths, they usually occur right after the strongest nectar flow seasons. So as long as your colony has the access and the ability to gather lots of nectar, hopefully food stores won't be an issue. If it does not, share food from other colonies or feed sugar syrup. Make sure the colony has access to water. Bees drown easily, so ensure that the water source has floats or something for your bees to land safely upon. Finally, be wary of robbing. This is the time of year when robbing can get

What Do Honey Bees and Penguins Have in Common?

Penguins and honey bees may not seem like they have much in common, but they do share one striking behavior: they both keep warm by clustering together as a group! In Antarctica, emperor penguins live in the most extreme conditions of any other warm-blooded animal on Earth. Antarctic winters include extreme temperatures with freezing winds of over one hundred miles per hour. Penguins survive these excruciating temperatures through a huddling behavior: they form a huddle of several thousand penguins with their heads tucked down. This tightly packed penguin huddle allows the group to preserve body heat. The penguins toward the center of the pack are protected from winds by those on the outer edges of the huddle. The penguins continually move from the outside of the huddle inward so that no emperor penguin is left to endure the harshest placement in the pack for too long. Honey bees perform a very similar function whenever the temperatures drop below 50°F. The honey bees form a tight cluster and move tiny muscles in their wings to generate heat. The queen stays at the center of the cluster, and the rest of the bees rotate so that no one is left at the coldest edges of the cluster for too long.

Of course, this is where the similarities end. In fact, emperor penguins and honey bees don't even exist on the same continent, as Antarctica is the only continent that does not have any bees at all! After bees, penguins are my favorite animal, and I actually visited Antarctica several years ago. Though I didn't get to see emperor penguins, I was fortunate to observe many other penguin species up close, and I can confirm firsthand that there are in fact no bees on the continent of Antarctica. (Efforts to convince my accountant that this was research and the trip was therefore work related and tax deductible, unfortunately, failed.)

started easily. Protect the hive by installing a robbing screen. You can purchase these or find plans to make your own online.

Review chapters 3 and 7 to refresh your memory on the nutritional needs of your bees and how to feed if stores are low in preparation for or during a dearth.

THE GROWING SEASON

The growing season is usually the season that most closely corresponds to spring in your area. The growing season has some of the year's first plants blooming and producing pollen and, eventually, nectar. This early pollen spurs the colony to start to "brood up," which is when the colony starts rapidly growing and producing more bees. Generally the growing season occurs just ahead of the nectar flow, as the colony wants as many adult foraging bees as possible to be ready to forage to collect nectar and make honey. Since it's three or more weeks before a newly emerged honey bee graduates to the foraging function, the queen must start building this workforce weeks ahead of this critical nectar flow period.

What to Expect

As the temperatures rise and the availability of pollen increases, a colony's population will begin to expand rapidly. More bees means more mouths to feed, and this is when those winter food stores become so important. Ironically, winter is not usually the time when I worry most about colony starvation: it's actually more of a concern during this brood buildup in the early part of the growing season. (This is, of course, presuming your colony entered into winter with plenty of food stores.) This is also the time of year when a colony needs more space for its growing brood nest. As the colony grows, it may begin to run out of space for brood, which can trigger swarming tendencies.

What to Do

The growing season will begin with a first post-cold weather dearth hive check. During these checks it is important to first fully assess the health of the colony:

- Is the queen laying eggs?
- Gauge the honey stores in the hive.
- Is the colony bringing in pollen to feed the new brood produced by the queen?
- Assess the population of each colony, especially in relation to one another.
- If a colony did not make it through the winter, be sure to clean up the hive, share any remaining honey with other colonies, and store the comb to prevent pests and preserve the beeswax comb to use in your other hives.

Review chapter 10 to refresh your memory of how to protect your comb from wax moth damage.

This is a great time to redistribute resources between colonies to help give each the best chance of quickly building through the growing season. Share honey, frames of stored pollen, or even frames of brood if necessary. However, be careful to not take too much brood from a colony that is just starting to grow its population after the winter dearth.

After this first inspection, it is important to continue to monitor colonies' food stores very closely. Starvation can happen quickly in the early part of the growing season when a colony is growing but nectar is not yet readily available. If food stores are not sufficient, either share honey from another hive or feed sugar syrup.

This is also the time of year when a colony is gearing up to try to swarm. Monitor a colony closely, and take steps to ease congestion and ensure that the queen has space to lay in the brood nest. You also may want to take advantage of your colonies' growth and look to make increases in your hive.

Reviewing chapters 8 and 9 on swarming bees and how to make splits will be helpful during the growing season.

NECTAR FLOW SEASON

Your area may have one or more nectar flows. Here in Central Texas, we usually get two: our strongest nectar flow occurs in the late spring months, May and into June, and we often see a smaller flow in the fall, usually in September or in October. This is what the bees are working toward all year: the opportunity to make honey! In most climates, bees actually make all of their honey in a very short period. Here in Central Texas, all the honey we harvest from a colony is usually produced in six weeks or less! Whether you have a good nectar flow, and your colony's ability to get out and gather that nectar, will determine whether they have enough food to make it through the dearth season. Of course, nectar flows can be wildly affected by any unusual weather patterns. This is because flower-blooming time is relatively very short for most plants. For example, a few years ago Austin had a wild week of rain every day in July, a time when our mesquite trees were in full bloom. The trees were producing nectar, but the bees couldn't forage in the rain, and when they did, the rain had washed away much of the nectar. As a result, the mesquite honey harvest was quite poor that year. A series of unfortunate weather events also can affect your nectar flow season. In the spring of 2022, we had the hottest May on record in Austin, and no rain at all. The flowers were blooming, but they simply weren't producing nectar without any rain.

What to Expect

During a nectar flow season, bees are bringing in incredible amounts of nectar very quickly. This is the time of year when they make the honey stores that must sustain them for the rest of the year and through the dearths. Populations tend to be at their highest, as the queen is continuing to respond to incoming resources by laying eggs heavily. However, once the nectar flow starts, colonies can quickly become what is called honey bound or nectar bound. This is when your foraging bees are bringing in nectar so quickly that they start filling the brood nest with nectar and honey, limiting the space that the queen has to lay her eggs. Honey-bound colonies may respond by swarming.

What to Do

In anticipation of the nectar flow and to help ensure colonies don't become honey bound, add supers to your strongest colonies! (Very weak or small colonies won't be able to make much, if any, honey.) If using top bar hives, this will entail moving your follower board back to allow the colony access to additional space to store honey. Avoid interfering in any way with a colony's ability to gather nectar and store honey.

Of course, it can be hard to predict when the nectar flows typically start in your area, and often they start much earlier or later than expected based on unusual weather patterns. Chatting with local beekeepers to learn when the nectar flows typically start in your area will be very helpful. Also be sure to take very detailed notes on when your first nectar flow starts and stops. Next year, these notes will help you better understand your area's flows. At the end of the nectar flow season, you may be able to harvest honey if any colonies produced excess honey!

Review chapter 11 to learn more about how to assess when your colony is on a nectar flow and steps to take, and to review how to harvest honey from your hive.

Lessons from a Beekeeper

Beekeeping, like other forms of agriculture, can be one of the most fulfilling experiences. It will also, at some point, certainly be one of the most heartbreaking.

On the evening of February 14, 2021, a once-in-a-lifetime kind of storm hit Texas. The temperatures and snow accumulation numbers I'm going to share won't seem that earth shattering to my friends in colder climates, but when you compare it to our typical weather during February and then place that into context of our infrastructure and systems and how unequipped we were to handle extreme winter weather, you will start to get a feel for how devastating this storm was.

Austin saw its fourth largest snowfall ever, and the most we've seen since 1949. Our average temperatures for February are 61 to 70°F. On February 11, 2021, our high temperatures fell below 35°F and dropped to single digits within a few days. We then experienced ten consec-utive days of temperatures at or below freezing. During these painfully record lows, more than 4.4 million people lost power for days, and some remained with-out electricity for weeks. Many more, 14.6 million, had disruptions to running water. Some experienced just low water pressure, but many, like us, lost water outright. Everyone in Austin was under a boil alert, and we boiled our water for days, some for weeks.

In my house, we personally lost power for the better part of a week, and within ten minutes of the power returning, our water access was cut off. We boiled snow to feed our animals, wash our dishes, and flush our toilets. We came within four gallons of having to drink boiled snow water. Oh, and did I mention that on the first day of losing power I learned I was pregnant? I couldn't confirm the results via a pregnancy test for four days because our roads were impassable.

For those of us who work in agriculture, the weather and resulting conse-quences of the storm were especially devastating. From farmers who didn't have water to feed their animals to those who lost entire acres of crops to the many folks who brought their backyard chickens indoors, it was weeks of survival mode for us all.

Unlike other forms of ag, however, there wasn't much beekeepers could do but hope and wait once the storm started. To be clear: bees are quite capable of dealing with the temperatures we experienced. Bees in cold weather climates are at these temperatures for months on end, and when these hard freezes hap-pen, a colony is usually broodless and can focus just on keeping the cluster and queen warm. However, the timing of this storm was devastating. Our growing season had already begun for bees, and nectar- and pollen-producing flowers had been in bloom for a few weeks. Our colonies were packed with brood. A colony forced to cluster to keep itself warm and also forced to try to protect developing brood is put at a severe disadvantage. Plus, Texas beekeepers, well into their growing season, had started hive checks, which meant propolis seals weren't as tight as they had been a month prior. To provide some context, I looked at my notes from the year prior. On this same week the year before, I was already out catching swarms!

This single weather event left us with losses six times the average loss rate of normal years and left most colonies too weak to split to make up for those winter losses. The weeks following the storm were devastating: cleaning up dead colonies, working to salvage honey off of hives before robbing wasps and bees could steal it, and working through how we would recoup those losses. In a year where we were still wrangling with the effects of the COVID-19 pandemic, it was almost too much to bear. And to top it off, the incredibly late freeze devastated the flora across the area, and for the first time in my beekeeping career, I was forced to feed my colonies pollen substitute because the plants that would normally bloom to provide that sustenance had died in the freeze.

Only two things are guaranteed in agriculture: a wild adventure full of surprises and a whole lot of heartache. Agriculture isn't for the faint of heart, but those with tender hearts do it best. And this storm proved that once again.

We are at the whim of Mother Nature in so many ways, and it's important that as beekeepers we understand how weather patterns are going to affect the flora and our bees. It's easy to get so focused on individual colonies that we can forget to step back and pay attention to the bigger picture. Prepare yourself now for the wins, and literal losses, of colonies that you will incur. We can't control the weather, but we can deepen our collective understanding of how weather dictates a colony's health. We can't predict the storms, but if we understand how these weather patterns will affect the flora, and ultimately our bees, we will become better stewards of our bees.

CHAPTER 13

Ensuring Beekeeping Success

Common Mistakes and Best Practices

An often repeated statistic is that 80 percent of beekeepers will quit within their first two years. I'm not certain where this stat originates from, and I don't know if it's true. I'm big on science-based facts, so I cringe a little whenever it's shared. That said, I've worked with and taught many beekeepers, and I have seen many get frustrated and throw in the towel after a short while. Though I can't say for sure that 80 percent of beekeepers quit within their first two years, anecdotal experience tells me that's probably not too far off, particularly for those who don't have an experienced beekeeper to help them along.

There are plenty of factors that deter people from sticking with it. A steep learning curve and the costs associated with the hobby, particularly if a beekeeper loses colonies each year, don't help. And, of course, bee stings aren't pleasant, and I see far too many beekeepers either not using their smoker or not wearing their gear correctly to help prevent those stings. Unrealistic expectations of honey coupled with a lack of understanding about the challenges of beekeeping can produce disappointing results. Emphasizing honey over the health of the colony, it turns out, isn't an effective way to meet any honey goals.

But beekeeping also has been one of the most rewarding activities in my life, and I know many of my students feel the same. How can I set you up for success?

Let's approach this from a tactical standpoint first. In no particular order, here are the top ten mistakes I see first-year beekeepers make again and again. We've covered all of these in previous chapters, but for your benefit I have provided a more detailed explanation of why these are so important.

1. Not Learning How to Spot Eggs

New beekeepers spend a lot of time and worry on activities that, overall, just don't provide that much information about what's going on in the hive. The biggest offender is queen spotting. Learning to identify a queen is very important, and there are a few occasions in your first year when you may need to find her. But for regular hive inspections, spotting a queen bee is of limited use: it tells you the colony has a queen. But what a beekeeper really needs to know is the *status* of that queen: is she laying eggs? Queens may slow or stop laying for myriad reasons, and if egg laying has ceased outside of the coldest months, you need to know about it! The only way to know for sure whether a queen is a laying queen is to look for eggs. Unfortunately, egg spotting is one of the most challenging activities for new beekeepers. Review the tips in chapter 5 to make sure you can spot eggs in your hive.

2. Not Feeding (or Overfeeding) New Colonies

Few issues can ignite a beekeeping war quite like feeding bees. I'm very squarely in the natural beekeeping camp, which means I only feed bees if their survival truly depends on supplemental feeding. I'm also a beekeeper with an open mind and understand that what is right for my bees may not be right for you and your bees. We all should spend time developing our own philosophy, which can vary based on our beliefs, our goals, and the amount of time and energy we think is right to spend with our bees. However, no matter your philosophy, feeding colonies that are not yet established is critical to their survival. What do I mean by an "established" colony? If your colony does not have, at a minimum, the equivalent of at least one deep hive body of drawn comb, your queen does not have the ability to lay eggs to produce the offspring necessary to care for and forage for the colony. Fortunately, sugar water in a ratio of one part sugar to one part water can provide the carbohydrates necessary for comb building. Aim to feed your new package or nuc one gallon of sugar water per week until it is established. Don't fall into the trap of believing that if one gallon is good, three gallons a week must be better. You can overfeed to disastrous consequences. Overfeeding can cause a colony to begin to store the sugar water in the comb, leaving the queen no space to lay eggs.

3. Not Using the Smoker Appropriately

Light a smoker every single time you open a hive or work within a few feet of the entrance. A smoker is critical to the safety of the beekeeper and the bees during a hive inspection. Bees communicate via pheromones, and the smoker masks the alarm pheromone given off by the guard bees at the entrance of a hive when they feel threatened. This same pheromone is released as a bee stings you.

A first-year colony is likely to be very docile, which may lead a beekeeper to believe they can easily work without a smoker and then get complacent and undervalue its importance. Without practice, they never learn how to properly light a fire or effectively use smoke with their hive. Once the colony is established and strong, the bees are going to be more inclined to protect themselves and their home, and new beekeepers can be caught off guard by a bad experience. It can be a traumatizing experience, and I've seen new beekeepers quit after taking a dozen stings because they never learned how to use the smoker.

4. Focusing Too Much on Harvesting Honey

If your only goal in beekeeping is to harvest honey, I encourage you to stop here and not spend another dime. You can purchase a lot of honey from a local beekeeper for the money you may invest in your first year as a beekeeper! If you keep bees for the honey, you are always going to be disappointed. If you keep bees for the experience, you'll never be disappointed. Beekeeping is a challenging, albeit very rewarding, endeavor. Beekeepers rely on live animals whose single food source depends on the weather to produce that food. Beekeeping is an activity full of factors we cannot control. Of course, nothing tastes as good as fresh honey from your own hive. However, the chances that a first-year colony will have enough resources to establish itself, store enough honey to feed the colony over the winter, and produce excess honey for you is unlikely. Therefore, push all the lessons about harvesting into your second year as a beekeeper. Spend the first year focusing on understanding how to keep your colony strong and healthy to ensure that it survives the winter months.

5. Focusing Too Much on Opportunistic Pests and Their Threat to the Colony

Small hive beetles (SHB) and wax moths are opportunistic pests. They take advantage and thrive when a colony has been weakened by other circumstances, such as going queenless or suffering from parasitic mite syndrome. Your best defenses against both of these pests is to ensure that your hive has at least three or four hours of solid sunlight every day and to focus on the activities necessary to keep your colony strong and healthy. Strong colonies can almost always patrol SHB and wax moths without intervention. It's only a colony weakened by another event that is most susceptible to the destruction they can cause. If you unwittingly assign a false narrative to the loss of a colony, you won't actually learn the real cause of the loss and are likely to repeat the mistakes that led to it.

6. Not Equalizing Resources between Colonies

Many different kinds of resources can be shared between colonies: all stages of brood, honey, nectar, pollen, bee bread, and drawn beeswax comb. When inspecting hives in an apiary, I make note of what each colony needs to help it thrive and what each colony can safely share with another colony in need. After an initial assessment of each colony's needs, I work back through the apiary, swapping beeswax frames of food and brood accordingly. A frame of capped brood will result in several thousand emerging bees in less than two weeks, which can help a colony with a low population. Similarly, sharing a frame of nectar or honey with a colony that lacks the carbohydrate stores to feed its population will do a lot more for the colony than a gallon of sugar water. But sharing resources isn't just a great solution for boosting weak or light colonies; it can also help prevent swarming. Don't allow one colony to grow so strong that it's on the verge of swarming while a second colony is so weak that it struggles to protect itself from opportunistic pests. Sharing frames of brood from a stronger to a weaker colony will help ease congestion and may prevent you from losing half of that strong colony to a swarm! If you keep equalizing resources at the top of your mind, your colonies will be much healthier for it.

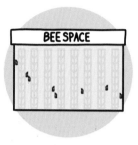

7. Not Managing Space Properly

One of the benefits of a beekeeper-managed colony is that we can add to or take away space as circumstances allow. Think of a colony as an accordion: it expands during the warmer spring months and contracts during the colder winter months. Beekeepers add space to strong colonies during the months when the queen is laying lots of eggs, the colony is expanding, and the bees are making honey. When preparing hives for winter, a beekeeper consolidates the space, as the colony will shrink considerably and a smaller space can help with temperature control. Further, when a colony's population shrinks because it has gone queenless or is weakened by disease, decreasing the space can help the colony better protect itself from pests. I've saved many colonies by simply moving them into a smaller space until they grow strong enough to better protect themselves.

Only give a colony as much space as it needs, and think about its health and your expectations of the colony in the next month. Is it a time of year when you expect the colony to be growing or shrinking? Is the colony queenless and not going to replenish worker bees while they work through the process of re-queening themselves? Does the colony have lots of capped brood, and therefore you expect the population to grow exponentially in the next ten days? Proper space management is a skill achieved with experience as you begin to learn the nuances between a "weak" and a "strong" colony and gain an understanding of your beekeeping calendar.

8. Not Thinking Enough about Varroa Management

We could dedicate all the pages in this book to varroa mite management and still barely scratch the surface of the topic. Varroa mites are external parasitic mites that attach and feed on the bees' fat bodies. Varroa mites can reproduce only in the brood cells of a bee hive. The mites weaken the bees, and the wounds left by the feeding become sites for various viruses. Effective varroa management starts with strong genetics: try to find a breeder who produces bees with the VSH trait. Learn about the tools that have been proven to help slow the spread of varroa. Chemical treatments do exist, but overuse has caused resistance and many are no longer effective. If you do decide to treat, use as a true treatment and not as a preventative. Follow the instructions of the treatment, as some cannot be used when brood and/or honey supers are on the hive. Also know that other nonchemical tools exist, including genetic control, brood breaks, requeening, screened bottom boards, and sanitation methods. Research your options and develop a plan. One very helpful resource is the Honey Bee Health Coalition. Check out their website for an assortment of helpful tools and information.

9. Not Taking Notes

Quick, tell me what you had for lunch three weeks ago today! You don't remember?

The same logic applies to beekeeping. Few can remember the necessary details of a hive inspection by the time the next hive inspection rolls around! After almost a decade of beekeeping, I still keep detailed notes on every colony from every hive inspection. Over time, my shorthand has gotten shorter and I have learned what information is most helpful to guide me in my decision-making. But I recommend beginning beekeepers take as many detailed notes as possible. Detailed notes will help capture the seasonality of beekeeping and will help you understand how the needs and actions of a colony change over the course of a year. Plus, it is helpful to have a note of important milestones, such as when the nectar flow started and stopped and when the worker bees removed the drones from a hive. There are several beekeeping note-taking applications available for your phone, though I prefer just freestyling my own notes. You also can use a pen and paper, use a notes phone app, or dictate to a voice-recording app. Choose whatever will best ensure that you take the notes.

Write the notes in a way that will help you prepare for the next inspection. For example, if a colony was very light at the last inspection, make note that you may need to bring sugar water. If you shared brood with a weaker colony, make note of that so you know you may need to share again. Then be sure to review these notes before you begin your next inspection. Detailed note-taking will maximize what you can learn from your colony over time.

10. Not Understanding How to Read Your Frames

You must learn how to read your frames and be able to identify what you are seeing. You cannot be an effective beekeeper if you cannot read the frames. I've described in detail how to identify all stages of brood, pollen, bee bread, nectar, honey, royal jelly, drone versus worker brood, and queen cells. But finding a local, experienced beekeeper who can confirm your identification skills is invaluable. If you are unable to find one willing to help, take close-up photos of your frames and use the internet to help confirm what you think you're seeing.

Make sure you understand what story the frames are telling. Close examination of your frames will reveal whether your brood is healthy or diseased, whether your colony is queenright (and if not, how long it has been queenless), whether a queenless colony has begun the requeening process, among many other things. The frames in a hive will tell you everything you need to know—you just have to understand how to read the story.

And finally . . . I want you to be successful. My heart breaks when I get calls from new beekeepers distraught over a colony loss. So with that, I'll leave you with my best tips to ensure that you aren't part of the 80 percent.

1. **Keep bees for the experience, not for the honey.** Honey isn't as easy to produce as some beehive marketing campaigns may lead you to believe. Plus, a secret: if you manage your bees in a way that puts their health first, your colonies are more likely to be productive in the area of honey. It's a win all around! If you keep bees for the experience, you will never be disappointed, but if you keep bees for the honey only, I can almost guarantee you disappointment.

2. **Touch as many hives as you can.** This means you need to look beyond your own apiary to gain more hands-on experience. Join a local club, hang out at your local beekeeping store, and make friends in the community and then work your bees together. If you have two hives and a friend has two hives and you work your bees together, by the end of the year you will have *double* the experience you would have had otherwise. Imagine the results if you made a third beekeeper friend!

3. **Acknowledge and accept that beekeeping is nuanced, regional, and very seasonal.** You can't learn all you need to know in a two-hour beginning class. Beekeeping is entering into a lifelong contract of learning, and that's what makes it fun and interesting. Enjoy the ride for what it is. We are all still learning, even those of us who have had bees for years.

4. **Find an experienced beekeeper and learn from them.** The best way to learn beekeeping is alongside an experienced beekeeper. Unfortunately, finding a trusted mentor can be challenging, depending on where you are. Lots of folks are looking for mentors, and there are only so many quality mentors with so much time to go around. Ideally, any mentor will have had bees for at least four or five years successfully. I define success here as someone whose losses are minimal. If a potential mentor loses their colonies every year and only keeps the hobby going by constantly buying new bees, look elsewhere.

   Also, keep in mind that you may need to pay for this expertise. The internet has a multitude of free resources, blogs, Facebook groups, and YouTube videos. But the internet is also full of folks who claim to be experts and who sadly have no idea what they are talking about. And it is full of folks who are experts but aren't necessarily great teachers. Teaching is an art and a skill set. Try to find a teacher you trust, and take a class or private instruction from them. A one-hour lesson in your hives with one of our beekeepers can give you more practical knowledge than an entire season's worth of online research. If you have someone you trust, invest in an hour with them. Many balk at the notion of paying for beekeeping education, arguing that it should be free. But all the same reasons you would reach out for help may make you willing to pay for demonstrated experience and knowledge: it is a skill set that takes many years to perfect. (Plus, I'd love to see you make that argument the next time you have to visit a lawyer's office!) You may find a mentor willing to dedicate their time and experience to you for free, but there are likely far more wannabe mentees than mentors in your area.

Reach out to your state apiary inspector's office too. Most states have at least one inspector, and they have a wealth of knowledge and information that you couldn't find elsewhere. For example, here in Texas the state inspector sends out notifications when the spraying of certain pesticides will happen, allowing the beekeepers in those areas to take action to help protect their bees.

That's it, y'all. I've not shared quite all I know, but I've shared all that I wish I had known before I jumped head-first into those two hives all those years ago. I hope this book empowers you but also humbles you as a student. I hope it makes you smile and wows you at every page turn. I hope I represent the bees well. And I hope this book motivates you to start your own apiary and, more importantly, inspires you to notice more of the smaller creatures, learn more about their importance, and understand how we are all connected. There's so much beauty to be seen, appreciated, and protected around us every day; we just have to look for it.

Photo taken outside Kabul,
Afghanistan, circa 2009. Author photo.

(ABOVE) Spring 2014, when I was terrified of bees. Author photo.
(BELOW) Photo taken in 2022 at the Honey Ranch, just east of Austin, Texas.
© 2022 by Noëlle Westcott.

Acknowledgments

This book has been an incredible work of labor that still feels as if it may never really come to fruition. (Even as I write this, I'm wondering if it will see the light of day.) Presuming that someday this does get printed and someone actually reads it, I want to take a moment to thank the many who had a hand in this book.

The origin story of this book started with Evie Carr, a cherished member of the Two Hives Honey team. Twice a year we offer an intensive, hands-on bee-keeping apprenticeship at the Honey Ranch, and the how-to book that we used as our text went out of print. Frustrated with the lack of books that focused on a holistic approach to beekeeping but were *also* a fun read, I entertained the idea of writing my own but quickly dismissed the idea. Mere days later, Evie offered the same suggestion, and it was just the push that I needed to suspend disbelief and naively jump head first into a delusional mosh pit.

Many thanks to the team at UT Press, specifically my editor Casey Kittrell and Gianna LaMorte. Gianna first heard my CIA-turned-beekeeper story from a chance meeting with another friend and approached me, wanting something more akin to a memoir. I will be forever grateful to her for letting me pitch her a beekeeping how-to book instead. I am especially grateful to Casey, who held my hand through the process and was incredibly patient with me as I tried to circumvent the process many times. Thanks to Jordan Price, that new friend who originally sung my praises to Gianna at a social event and who would later entrust me to bring fifteen beekeepers on a "beekeeping safari" to his beloved Malawi.

Thank you to those who patiently and graciously read this manuscript and offered notes and suggestions: my UT Press peer reviewers, Casey Kittrell, Adrienne Dawson (who worked on this while home sick with COVID-19), Celia Bell, Evie Carr, and Vicki Blachman. Though I perhaps didn't always sound grateful for the suggestions, I am, and you made this an exceptionally better manuscript as a result. Thanks to the good people at Texas A&M University and the Bee Informed Partnership who helped with photos for the book.

Thank you to the many beek apprentices over the years who read parts of this manuscript as part of their beekeeping curriculum and overlooked the many typos, grammatical errors, and just plain gibberish of those early versions. I know with each passing class I promised you would be the one that actually got a book in your hands. Someday I will promise this and it won't be a lie. My desire to give you the best out there is what got us here. I hope I made you proud.

Thank you to the entire Two Hives Honey team. Your support and belief in the mission that we can do big things by acknowledging the smallest creatures in life keeps me going. I can't thank you enough for continuing to believe in the work we do daily.

To Caroline Brown, who provided the illustrations in this book: thank you for sharing your talents with us. What started as a sort of Instagram competition to find a muralist for the Honey Ranch has ended up as a beautiful friendship. You and your illustrations are the reason for the joy found in this book that makes it like no other. No one else will understand how important you were in helping this vision come to life, but I hope you know I do.

Thank you to my incredibly solid support system: my mother and my husband, Aidan, who both believe everything is possible, as long as I am at the helm. I don't know what I did to dominate in the unwavering support category of life, but I do and I am so grateful. Aidan: you married a woman who has dreams far bigger than one lifetime can possibly allow, and you have watched me self-destruct countless times in my efforts to try to achieve them anyway. Your ability to still support every one, even though they often come at a personal cost to you, is incredible. You deserve so much more than a thank you, and I hope that someday I can come up with the words to do your love and support justice.

And finally to my little Atlas, who as I write this is sleeping in his crib with the only stuffed animal he has ever shown affection for: a bee. I wrote almost half of this book in the two months that I had you in my belly but no idea you were there. After I discovered I was pregnant, it took me another eighteen months to finish the second half of the book. I finished the first draft of this book in your first year of life, turning it in two weeks after your first birthday. That I believed writing a book while raising a newborn was even possible demonstrates my naivete about this whole process. But I kept going, wanting to prove to you and me that we can still do big things together (just a little bit more slowly).

And of course, all this would not be if it were not for the bees. My greatest hope is that this book introduces so many more to the mind-blowing, often frustrating, astonishing, bewildering, sometimes disconcerting, but overall extraordinary experience of beekeeping.

Glossary

abdomen: A part of a honey bee's anatomy that houses the primary organs for digestion and stinging.

absconsion: When a colony, including a queen bee, decides to abandon its home and look for a new place to nest.

age polyethism: The division of specialized labor based on the age of a social insect. Worker honey bees exhibit age polyethism when they perform different activities at different times of their lives.

apiary: A collection of or area where one or more honey bee hives are kept.

bee bread: Pollen that has undergone a fermentation process by the worker bees and used to feed developing larvae.

bee space: Describes the gaps that honey bees will not fill with beeswax comb or propolis, allowing for free movement of the bees around the hive. Lorenzo Langstroth discovered that bees will fill any space less than ¼ inch with propolis and will build beeswax comb in any space ⅜ inch or larger.

beeswax: A product produced by worker honey bees from glands on their abdomens. Honey bees shape the beeswax into hexagons as a place to house food, water, and developing honey bees.

brood: The young of an animal. For honey bees, we use the term "brood" to collectively refer to any or all of their eggs, larvae, and pupae.

capped brood: The pupal stage of a honey bee. Capped brood is named after the beeswax capping covering the developing pupae.

carbohydrate: A macronutrient that provides fuel and energy for the honey bee colony to function. The primary sources of carbohydrates for the colony are both nectar and honey.

colony: A distinguishable population of honey bees that live and work together to ensure survival.

dearth: Periods when there is little to no food available for honey bees because the climate does not support the blooming of pollen- and nectar-producing plants. Dearths can occur in extreme cold, hot, or dry climates.

drawn comb: Sheets of connecting hexagons created by bees from beeswax.

drone bee: A male member of a honey bee colony. Drones bees are responsible for mating with queen bees.

drone congregation areas (DCAs): Areas where drones gather to wait for virgin queens to arrive for mating.

egg: The first stage of the honey bee development cycle.

endophallus: A part of a male drone bee's anatomy critical to the function of mating and fertilizing a queen bee.

eusocial: A term used to describe a species that has cooperative brood care, overlapping generations, and a reproductive division of labor that includes both sterile members and reproductive members.

exoskeleton: The external skeleton that supports and protects the honey bee body.

forager bee: One of the oldest worker bees in a colony. Forager bees are responsible for gathering pollen, nectar, water, and resins to make into propolis.

head: A part of a honey bee's anatomy that contains the eyes and mouthparts, along with two antennae.

honey: A product produced by honey bees by dehydrating, or curing, flower nectar to 18.5 percent water or less.

honey-bound colony: A colony in which bees start filling the brood nest with nectar, honey, or sugar syrup, preventing the queen from laying brood.

honey crop: A part of a honey bee's anatomy responsible for carrying nectar back to the hive.

honey flow: The period of time that bees have access to enough nectar that they can begin to store it as honey. Also called a nectar flow.

hypopharyngeal glands: The part of a worker bee's anatomy that runs along the side of the head and produces royal jelly.

integrated pest management (IPM): A strategy for managing pests and their damage through a combination of techniques that minimize the risk to people, property, resources, and the environment. In an IPM strategy, pesticides are used *only* after monitoring indicates their need and only according to established guidelines.

Langstroth hive: A type of modular honey bee hive that consists of a series of stacked boxes known as hive bodies. Also often called Langs by beekeepers.

larva: The second stage of the honey bee development cycle.

lipid: A macronutrient that includes fats and essential fatty acids such as omega-3 and omega-6 in a honey bee colony. Honey bees obtain lipids from pollen.

long Langstroth: A type of hive situated along a horizontal access that uses standard deep frames built for a Langstroth hive. Often referred to as a long Lang.

macronutrient: Nutrients derived from food sources that make up the largest portion of an organism's diet.

mandibles: A part of a honey bee's anatomy made of two jaws that swing in and out from the bee's head. The mandibles are used for all kinds of household functions, like building comb and feeding the developing honey bees.

micronutrient: Vitamins and minerals that are necessary in much smaller quantities than macronutrients.

nectar: A sugar-rich liquid produced by some flowers in order to attract pollinators such as honey bees, butterflies, and hummingbirds to help ensure pollination of the plant. Honey bees gather nectar to feed their colony.

nectar flow: The period of time that bees have access to enough nectar that they can begin to store it as honey. Also called a honey flow.

nucleus colony: A small colony of honey bees containing frames or top bars of beeswax comb, brood, honey and pollen, and a laying queen bee. A nucleus colony can be used by beekeepers to start a new honey bee hive. Also called a nuc.

nurse bee: A young worker bee responsible for caring for the developing bees that have yet to reach adulthood in a colony.

open brood: The eggs and/or larval stage of a honey bee. Open brood gets its name from the fact that it is void of the beeswax capping that is placed over the pupal stage of the honey bee. Beekeepers also call this uncapped brood.

package of bees: A ventilated vessel containing worker bees and a queen bee, used by beekeepers to start a new honey bee colony. Packages of bees are usually measured and sold by weight.

pheromone: A chemical substance that allows honey bees to send messages via odor. Honey bees use pheromones in many ways, including to warn the colony of danger, communicate where to find food, and ensure that their queen is alive and well.

pollen: A substance discharged from the male part of a plant that is necessary for fertilizing female plants. Pollen is gathered by honey bees and is important in brood rearing.

pollen baskets: A part of a worker honey bee's anatomy located on her hind leg that is used to collect and transport pollen.

proboscis: A tube-like structure that acts like a straw and works as the bee's tongue.

propolis: Produced by honey bees by mixing resin gathered from some botanical sources, saliva, beeswax, and honey or nectar. Honey bees use the propolis to weatherproof their hives and help protect the colony from pathogens. Beekeepers also call it bee glue.

protein: A macronutrient necessary for muscle and glandular development and for repairing tissues in a honey bee colony. The primary source of protein for a colony is pollen.

pupa: The third stage of the honey bee development cycle.

queen bee: The fertile female of a honey bee colony. Generally speaking, colonies have just one queen bee, though there are some exceptions of large colonies with two or more queens.

queen cage: A small screened box used by beekeepers to house a queen bee.

queen cell: A vessel that worker bees create out of beeswax to house developing queen bees.

queenless colony: A colony that does not have a living queen bee.

queen mandibular pheromone (QMP): A pheromone produced by the queen bee that regulates worker bee activities in a colony, attracts a retinue, promotes clustering behavior, and helps attract potential drone mates.

queenright colony: A colony that has a living queen bee.

retinue: The worker honey bees in a colony responsible for cleaning, feeding, and tending to the queen bee's every need.

royal jelly: Protein-rich secretions produced by glands on the heads of the worker nurse bees.

scout bee: An older worker bee in charge of seeking out a new home when it's time for the colony to swarm, or reproduce.

superorganism: An organized society of social organisms that function as an organic whole. Honey bees, ants, and termites are all considered superorganisms.

swarming: The means by which a honey bee colony reproduces. The honey bees and queen leaving the original colony in this form of reproduction is called a swarm.

thorax: The part of a honey bee's anatomy responsible for locomotion. Four wings and six legs are attached to the thorax.

top bar hive: A type of hive composed of a long, singular box. Inside the hive, bees attach their beeswax comb to top bars, bars that sit across the top of the box.

uncapped brood: The eggs and/or larval stage of a honey bee. Uncapped brood gets its name from the fact that it is void of the beeswax capping that is placed over the pupal stage of the honey bee. Beekeepers also call this open brood.

varroa sensitive hygiene (VSH): A desirable behavioral trait that allows honey bees to detect and remove pupae that are infested with varroa mites.

waggle dance: A movement honey bees use to communicate in a colony. Honey bees use the waggle dance to communicate the location of food and potential nesting sites.

Warre hive: A type of hive composed of a modular design with a series of vertically stacking boxes.

worker bee: An infertile female of a honey bee colony. Worker bees are responsible for all functions of a colony except for reproduction, including gathering food and water, cleaning the hive, heating and cooling the hive, and caring for the young.

Notes

CHAPTER 1: HONEY BEE BIOLOGY

1. Though honey bees are not native to North America, it's important to note that we do have roughly four thousand species of native bees. Most of these are solitary bees, meaning they do not live in large colonies like honey bees. However, species of bumble bees, one type of native bee, do live in small colonies of two to three hundred bees. Some other examples of native bees include mason bees, carpenter bees, sweat bees, leaf cutter bees, and mining bees. Most of the bees native to North America are ground nesters.

2. Michael Simone-Finstrom et al., "Propolis Counteracts Some Threats to Honey Bee Health," *Insects* 8, no. 2 (April 29, 2017): 46, https://doi.org/10.3390/insects8020046.

3. Freddie-Jeanne Richard et al., "Effects of Insemination Quantity on Honey Bee Queen Physiology," *PloS One* 2, no. 10 (October 3, 2007): e980, https://doi.org/10.1371/journal .pone.0000980.

4. Though most colonies have just one queen bee, there are some exceptions to this rule. Not a lot is known about when and how a colony may have two queens, but I have observed very large wild hives with more than one queen, almost as if two colonies were coexisting together. I have also observed the phenomenon of more than one queen a handful of times in my own hives, and I suspect it happens far more often than we realize. Because we all accept the assumption that every colony always has one queen, we stop looking as soon as we see the queen. Unless the two queens happened to end up beside one another on the frame, we would never know that a second queen may be living in the colony! Theories about how and why some queens allow a daughter queen to survive include that it could be a mother-daughter coexisting to produce a sort of "mega-colony" or that having two queens simply means the colony has reared a replacement for the older mother but hasn't killed her off yet. We can't be sure, but I chalk it up to one of the many inexplicable phenomena in honey bee colonies.

5. The geek in me can't help but expand on this further. The artificial banana flavor similar to that of the alarm pheromone wasn't actually developed based on the variety of banana we find at the grocery store today. Reportedly, it was based on a variety of banana called the Gros Michel, which was widely available in the United States until a fungus wiped out the species in the 1950s. So perhaps the advice shouldn't be to avoid eating bananas near a hive but rather to avoid banana flavored candy!

CHAPTER 2: A HONEY BEE'S HOME

1. It is rare, yet not unheard of, for a honey bee colony to build what is known as an open-air hive. An open-air hive is one that is built in a location exposed to the elements and not inside another protected vessel. I've most often seen these hanging from branches under a canopy of trees. Open-air hives are fascinating and quite beautiful, and they provide unique opportunities to observe a colony's activities. Of course, these colonies are also the most exposed to predators and rain, snow, and extreme temperatures. If you stumble across one, enjoy the rarity of your find!

CHAPTER 3: HONEY BEE NUTRITION

1. Different sources report different foraging radiuses, though the most common distances I find reported are three to five miles. The British Columbia Ministry of Agriculture, Food and Fisheries reports that bees are known to fly as far as 12 km (7.4 miles), but foraging is limited to food sources within 3 km (1.86 miles) and 75 percent of foraging bees will typically fly no more than 1 km (0.62 miles) from the hive. *Apiculture Factsheet #111: Bee Behavior during Foraging,* November 2020, https://www2.gov .bc.ca/assets/gov/farming-natural-resources-and-industry/agriculture-and-seafood /animal-and-crops/animal-production/bee-assets/api_fs111.pdf.

2. Amanda Ellis et al., "The Benefits of Pollen to Honey Bees," University of Florida Institute of Food and Agricultural Sciences, November 2020, https://edis.ifas.ufl.edu /publication/IN868.

CHAPTER 7: FEEDING HONEY BEES

1. It's worth noting that honey also produces HMF when heated to higher temperatures for extended periods.

CHAPTER 9: GROWING AN APIARY

1. J. Rhodes, *Drone Honey Bees—Rearing and Maintenance* (New South Wales Department of Agriculture, 2002), https://www.dpi.nsw.gov.au/__data/assets/pdf_file/0007 /117439/drone-bee-rearing-and-maintenance.pdf.

CHAPTER 10: PARASITES, PATHOGENS, AND PESTS

1. Isabel Schödl, et al. "Simulation of Varroa Mite Control in Honey Bee Colonies without Synthetic Acaricides: Demonstration of Good Beekeeping Practice for Germany in the BEEHAVE Model," *Ecology and Evolution* 12, no. 11 (November 2022): e9456, https://doi.org/10.1002/ece3.9456.

2. Arne Kablau et al., "Hyperthermia Treatment Can Kill Immature and Adult Varroa Destructor Mites without Reducing Drone Fertility," *Apidologie*, no. 51 (2020): 307–315, https://doi.org/10.1007/s13592-019-00715-7.

3. Scott Weybright, "Fungus Fights Mites That Harm Honeybees," Washington State University, news release, May 27, 2021, https://news.wsu.edu/press-release/2021/05/27 /fungus-fights-mites-harm-honey-bees.

4. Paul E. Stamets et al., "Extracts of Polypore Mushroom Mycelia Reduce Viruses in Honey Bees," *Scientific Reports*, no. 8 (2018): 13936, https://doi.org/10.1038/s41598-018-32194-8.

5. Honey Bee Health Coalition. www.honeybeehealthcoalition.org.

6. In 2017 the US Food and Drug Administration amended animal drug regulations; they now require beekeepers to seek a prescription in order to obtain antibiotics to treat EFB.

7. Business Wire, "First-in-Class Honeybee Vaccine Receives Conditional License from the USDA Center for Veterinary Biologics," news release, January 4, 2023, https://www.businesswire.com/news/home/20230104005262/en/First-in-Class-Honeybee-Vaccine-Receives-Conditional-License-from-the-USDA-Center-for-Veterinary-Biologics.

8. In 2020 the genus *Nosema* was reclassified as *Vairomorpha*. You may continue to see the old genus *Nosema* or the new classification of *Vairomorpha* in your additional readings.

CHAPTER 11: HARVESTING FROM THE HIVE

1. Nermeen Yosri et al., "Anti-Viral and Immunomodulatory Properties of Propolis: Chemical Diversity, Pharmacological Properties, Preclinical and Clinical Applications, and In Silico Potential against SARS-CoV-2," *Foods* 10, no. 8 (July 31, 2021): 1776, https://doi.org/10.3390/foods10081776.

2. United States Food and Drug Administration Center for Food Safety and Applied Nutrition, *Proper Labeling of Honey and Honey Products: Guidance for Industry*, February 2018, https://www.federalregister.gov/documents/2018/03/02/2018-04282/proper-labeling-of-honey-and-honey-products-guidance-for-industry-availability.

Suggested Resources

This is a sampling of books, journal articles, studies, blogs, magazines, and podcasts that I have found useful in my own honey bee education. Remember that beekeeping is full of many perspectives and management techniques. A wide variety of points of view, such as that represented by the following list, can be useful in forming your own philosophy.

American Bee Journal. www.americanbeejournal.com.

Arien, Yael, Arnon Dag, Shiran Yona, Zipora Tietel, Taly Lapidot Cohen, and Sharoni Shafir. "Effect of Diet Lipids and Omega-6:3 Ratio on Honey Bee Brood Development, Adult Survival and Body Composition." *Journal of Insect Physiology*, no. 124 (2020): 104074.

Bee Culture Magazine. www.beeculture.com.

Bortolotti, Laura, and Cecilia Costa. "Chemical Communication in the Honey Bee Society." In *Neurobiology of Chemical Communication*, edited by Carla Mucignat-Caretta. Boca Raton, FL: CRC Press/Taylor & Francis, 2014. https://www.ncbi.nlm.nih.gov/books/NBK200983.

Burlew, Rusty. Honey Bee Suite. www.honeybeesuite.com.

Canadian Association of Professional Apiculturists. *Honey Bee Diseases and Pests.* Edited by Stephen F. Pernal and Heather Clay. 3rd ed. 2013.

Cook, Daniel, Alethea Blackler, James McGree, and Caroline Hauxwell. "Thermal Impacts of Apicultural Practice and Products on the Honey Bee Colony." *Journal of Economic Entomology* 114, no. 2 (April 2021): 538–546.

Ellis, Amanda, Jamie D. Ellis, Michael K. O'Malley, and Catherine M. Zettel Nalen. "The Benefits of Pollen to Honey Bees." University of Florida Institute of Food and Agricultural Sciences. November 2020. https://edis.ifas.ufl.edu/publication/IN868.

Ellis, Jamie, and Amy Vu, hosts. *Two Bees in a Podcast.* Podcast. Produced by University of Florida's Honey Bee Research and Extension Laboratory. https://entnemdept.ufl.edu/honey-bee/podcast.

Flottum, Kim, and Jim Tew, hosts. *Honey Bee Obscura.* Podcast. https://www.honeybeeobscura.com.

Honey Bee Health Coalition. https://honeybeehealthcoalition.org.

Richard, Freddie-Jeanne, David R. Tarpy, and Christina M. Grozinger. "Effects of Insemination Quantity on Honey Bee Queen Physiology." *PloS One* 2, no. 10 (October 3, 2007): e980. https://doi.org/10.1371/journal.pone.0000980.

Sammataro, Diana, and Alphonse Avitabile. *The Beekeeper's Handbook.* 5th ed. Simon & Schuster Books for Young Readers, 2021.

Seeley, Thomas D. *Honeybee Democracy.* Princeton, NJ: Princeton University Press, 2010.

Seeley, Thomas D. *Lives of Bees: The Untold Story of the Honey Bee in the Wild.* Princeton, NJ: Princeton University Press, 2019.

Simone-Finstrom, Michael, Renata S. Borba, Michael Wilson, and Marla Spivak. "Propolis Counteracts Some Threats to Honey Bee Health." *Insects* 8, no. 2 (April 29, 2017): 46. https://doi.org/10.3390/insects8020046.

Simopoulos, Artemis P. "The Importance of the Omega-6/omega-3 Fatty Acid Ratio in Cardiovascular Disease and Other Chronic Diseases." *Experimental Biology and Medicine* 233, no. 6 (2008): 674–688. https://doi.org/10.3181/0711-MR-311.

Somerville, Doug. *Fat Bees, Skinny Bees: A Manual on Honey Bee Nutrition for Beekeepers.* RIRDC publication no. 05/054. New South Wales Department of Primary Industries, 2005.

United States Food and Drug Administration Center for Food Safety and Applied Nutrition. *Proper Labeling of Honey and Honey Products: Guidance for Industry.* February 2018. https://www.federalregister.gov/documents/2018/03/02/2018-04282/proper -labeling-of-honey-and-honey-products-guidance-for-industry-availability.

Yosri, Nermeen, Aida A. Abd El-Wahed, Reem Ghonaim, Omar M. Khattab, Aya Sabry, et al. "Anti-Viral and Immunomodulatory Properties of Propolis: Chemical Diversity, Pharmacological Properties, Preclinical and Clinical Applications, and In Silico Potential against SARS-CoV-2." *Foods* 10, no. 8 (July 31, 2021): 1776. https://doi.org/10.3390 /foods10081776.

Index

Note: Page numbers in italics indicate illustrations.